Communication Network Systems and Security

Communication Network Systems and Security

Edited by **Pascal Formann**

CLANRYE INTERNATIONAL

New Jersey

Published by Clanrye International,
55 Van Reypen Street,
Jersey City, NJ 07306, USA
www.clanryeinternational.com

Communication Network Systems and Security
Edited by Pascal Formann

International Standard Book Number: 978-1-63240-547-0 (Hardback)

Printed in the United States of America.

Contents

Preface

Communication network systems refer to the collection of various transmission networks, data terminal equipment (DTE), communication networks and tributary stations, etc. which are able to hold interoperation and interconnection to work as a single entity. Communication security enables these networks to work without hindrance and security failures. This book provides comprehensive insights into this field. The various studies that are constantly contributing towards advancing technologies and evolution of this field are examined in detail in this text. It is appropriate for students seeking in-depth knowledge about this subject, as well as for experts. This book will serve as a valuable guide for all associated with this field.

Significant researches are present in this book. Intensive efforts have been employed by authors to make this book an outstanding discourse. This book contains the enlightening chapters which have been written on the basis of significant researches done by the experts.

Finally, I would also like to thank all the members involved in this book for being a team and meeting all the deadlines for the submission of their respective works. I would also like to thank my friends and family for being supportive in my efforts.

Editor

Control Access Point of Devices for Delay Reduction in WBAN Systems with CSMA/CA

Akinobu Nemoto[1], Pham Thanh Hiep[2], Ryuji Kohno[2]

[1]Department of Medical Informatics, Yokohama City University, Yokohama, Japan
[2]Division of Physics, Electrical and Computer Engineering, Yokohama National University, Yokohama, Japan
Email: anemoto@med.yokohama-cu.ac.jp, hiep@ynu.ac.jp, kohno@ynu.ac.jp

Abstract

Due to the gathering of sickrooms and consultation rooms in almost all hospitals, the performance of wireless devices system is deteriorated by the increase of collision probability and waiting time. In order to improve the performance of wireless devices system, relay is added to control the access point and then the access of devices is distributed. The concentration of access point is avoided and then the performance of system is expected to be improved. The discrete time Markov chain (DTMC) is proposed to calculate the access probability of devices in a duration time slot. The collision probability, throughput, delay, bandwidth and so on are theoretically calculated based on the standard IEEE802.15.6 and the performance of the system with and without relay is compared. The numerical result indicates that the performance of the system with control access point is higher than that of the system without control access point when the number of devices and/or packet arrive rate are high. However, the system with control access point is more complicated. It is the trade-off between the performance and the complication.

Keywords

Standard IEEE802.15.6, Discrete Time Markov Chain Method, Control Access Point, Bandwidth Efficiency, Delay

1. Introduction

1.1. The Problem of WLAN in Hospitals

In almost hospitals, sickrooms and consultation rooms are respectively gathered at one place for convenience of patients. It may be good for patients and hospital sites, however, on the view point of wireless system, there is a problem. Since medical devices access the wireless local area network (WLAN) base station via wireless channel, the collision when more than one devices access the channel in the same time, will occurs depending on the number of devices and the number of data packets that be generated by every device in one second. Moreover, a lot of devices access the WLAN base station that is close to the consultation rooms (Wireless LAN 2 in **Figure 1**), whereas a few devices access the WLAN base station that is far from consultation rooms (Wireless LAN 1). The access of devices concentrates at Wireless LAN 2, consequently, the probability of collision increases, and then the throughput decreases, the delay increase. As a result the bandwidth efficiency decreases.

1.2. Aims and Motivations

Since many body functions are traditionally monitored and separated by a considerable period of time, it is hard for doctors to know what is really happening. This is the reason why the monitoring of movement and all body functions in daily life are essential. The delay of patients' data as well as the collision of data packets may let doctors misunderstand and information data be lost by timeout. In order to decrease the delay and increase the throughput, the relay can be set to avoid the concentration of WLAN base station. As shown in **Figure 1**), some devices assess the wireless LAN 1 via the relay, therefore, the number of devices that access the wireless LAN 2 is reduced, and then the bandwidth efficiency is expected to be higher. However, the delay due to signal processing at relay should be considered. At scheme 1, all devices access the wireless LAN 2, whereas at scheme 2, the relay is set and devices access the channel via either wireless LAN 1 or 2. The performance of both schemes 1 and 2 is mathematically analyzed base on standard IEEE802.15.6. The throughput, delay and bandwidth efficiency of both schemes are numerically compared.

1.3. Related Works

According to an emergency of wireless body area network (WBAN), the standard IEEE802.15.6 was established in Feb. 2012 [1]. An overview of the standard and performance analyses of WBAN based on bandwidth efficiency and delay were represented in [2]-[4]. In these papers, however, the WBAN is assumed to consists of only one device that keeps transmitting a data packet. Packet arrival rates and collisions due to transmission of multiple devices in the same time weren't considered. On the other hand, a Physical layer (PHY), Media Access Control (MAC) layer and network layer of WBAN were researched in [5] [6]. Furthermore, the control on MAC layer was analyzed to improve the performance of WBANs [7] [8]. The transmission of implanted devices was considered under conditions of low transmit power and low harmful influence on a human body [9] [10]. The performance of WBANs that has multiple devices and multiple user priorities were analyzed in both saturation [11] [13] [14] and non-saturation [12]. Additionally, WBANs were analyzed in further detail when a superframe with beacon mode and an access phases length were taken into consideration in [13] [14], respectively. However, efficiencies of number of devices, packet arrival rates, packet sizes, etc. on the throughput of each device and the total throughput, the delay and the bandwidth efficiency of system hasn't been discussed.

1.4. Organization of the Paper

The rest of paper is organized as follows. We introduce a brief of PHY and MAC layers of standard IEEE802.15.6

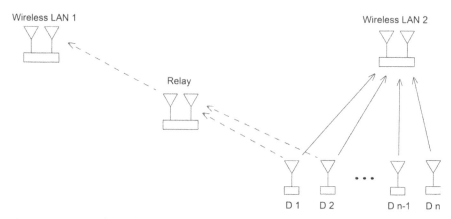

Figure 1. The wireless LAN system in a hospital.

in Section 2. The discrete time Markov chain is proposed and then the performance of both schemes 1 and 2 with CSMA/CA is analyzed in Section 3. The numerical evaluation of both schemes is described and compared in Section 4. Finally, Section 5 concludes the paper.

2. Brief of Standard IEEE802.15.6

A brief of the standard that related to our research is described in this section. The further detail of standard can be found in [1] [2].

2.1. PHY Layer

The IEEE802.15.6 defines three different PHYs, *i.e.*, human body communication (HBC), narrowband (NB) and ultra wideband (UWB). Furthermore, the NB is divided in several frequency bands and a data rate, symbol rate, etc. of every frequency band are different. We analyze the system in 2400 MHz - 2483.5 MHz band as an example, the analysis in different frequency band is similar. The physical protocol data unit (PPDU) of NB PHY is described in **Figure 2**. Components of PPDU are fixed, excepted the payload. Parameters of PHY layer are summarized in **Table 1**.

2.2. MAC Layer

The algorithm of CSMA/CA based on IEEE802.15.6 is described as follows. All devices set their backoff counter

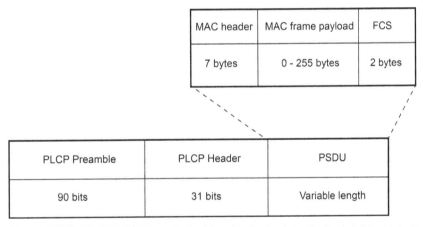

Figure 2. PPDU of NB PHY.

Table 1. Parameters of PHY layer.

Frequency band [MHz]	2400 - 2483.5
Packet component	PSDU
Modulation DBPSK	DBPSK
Symbol rate Rs [ksps]	600
Data rate $Rhdr$ [kbps]	242.9
Clear channel assessment [bits]	63
MAC header [bits]	56
MAC footer [bits]	16
CSMA slot time Ts [µs]	125
Short interframe spacing time T_{pSIFS} [µs]	75
Preamble [bits]	88
Propagation delay α [µs]	1

to a random integer number uniformly distributed over the interval $[1, W]$, where W is a contention window within $(W_{\min}; W_{\max})$. The value of $W_{\min}$ and $W_{\max}$ varies depending on the user priorities (UPs). However, in this paper, the UP of all devices is assumed to be the same as zero-th UP. The extension for multiple UPs is straightforward.

As shown in **Figure 3**, a device starts decrementing its back off counter by one for each idle CSMA slot. When the back off counter reaches zero, the device transmits its packet. Once the channel is busy because of transmission of another device, the device locks the back off counter until the channel is idle. The transmission is failed if the device fails to receive an acknowledgement (ACK) due to a collision or being unable to decode. The W is doubled for even number of failures until it reaches $W_{\max}$. The maximum number of back off stages is bound by a retry limit m. Once the number of retries exceeds the predefined retry limit m, the packet is discarded. When the transmission is successful, the W is set to $W_{\max}$. The W of zero-th UP is represented in **Table 2**.

3. Performance Analysis of WBANs

3.1. Discrete Time Markov Chain

At first, the performance of scheme 1 is analyzed. The scheme 1 consists of a single base station, the wireless LAN 2, and n devices in a star topology, $D1$, $D2$, $\cdots$, Dn (**Figure 1**). All devices can access the wireless LAN 2 directly, however, the wireless LAN 1 is out of them range. The discrete time Markov chain (DTMC) is proposed to calculate the access probability of each device in every time slot. The proposal DTMC of device i with empty state is described in **Figure 4** and notations used in this section are listed in **Table 3**. A packet arrival rate of all devices is assumed the same and denoted by λ. Hence, $\rho = 1 - e^{-\lambda Ts}$, where e denotes the Napier's constant, denotes the probability that the device has a packet to transmit in duration time of Ts. The transmission failed probability and the idle probability of device I are respectively expressed as

$$P_{i,\text{fail}} = P_{i,\text{col}} + \text{PER}_i,$$

$$P_{i,\text{idle}} = \frac{\prod_{k=1}^{n}(1-\tau_k)}{1-\tau_i}, \tag{1}$$

here $P_{i,\text{col}} = 1 - \prod_{k \neq i}^{n}(1-\tau_k)$. The state transmission probabilities of DTMC method are represented as follows.

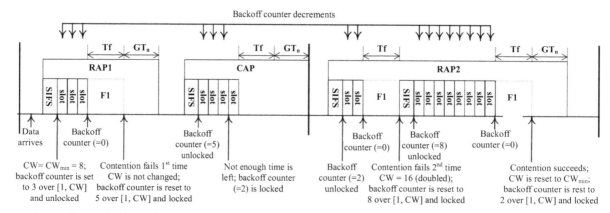

Figure 3. An example of operation of CSMA/CA and relationships of time durations.

Table 2. Contention window for every UP.

Number of retransmissions	0	1	2	3	4 and over
W	16	16	32	32	64

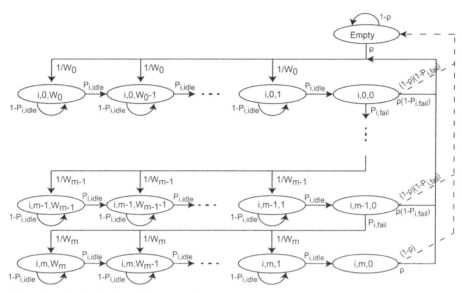

Figure 4. Algorithm of DTMC method.

Table 3. Explanation of notations.

Notation	Explanation
λ	Packet arrival rate during a unit time
ρ	Packet arrival rate during a slot time
m	Packet retry limit
n	Total number of devices
x	Payload size
$Total\ x$	Total data
T_i	Access probability during a slot time
$b_{i,k,j}$	Stationary distribution with backoff stage k, backoff counter j
$P_{i,\text{idle}}$	Channel idle probability
$P_{i,\text{fail}}$	Transmission failed probability
$P_{i,\text{col}}$	Collision probability
PER_i	Packet error rate
W_k	Contention window of k backoff stage

$$\Pr\{i,k,j|i,k,j+1\} = P_{i,\text{idle}}, \quad \text{for} \quad k \in [0,m], j \in [0,W_k],$$

$$\Pr\{i,k,j|i,k-1,0\} = \frac{P_{i,\text{fail}}}{W_k}, \quad \text{for} \quad k \in [1,m], j \in [1,W_k],$$

$$\Pr\{i,k,j|i,k,j\} = 1 - P_{i,\text{idle}}, \quad \text{for} \quad k \in [0,m], j \in [1,W_k],$$

$$\Pr\{i,0,j|i,m,0\} = \frac{\rho}{W_0}, \quad \text{for} \quad j \in [1,W_0],$$

$$\Pr\{i,0,j|\text{empty}\} = \frac{\rho}{W_0}, \quad \text{for} \quad j \in [1,W_0],$$ (2)

$$\Pr\{\text{empty}|i,k,0\} = (1-\rho)(1-P_{i,\text{fail}}), \quad \text{for} \quad k \in [0,m-1],$$

$$\Pr\{\text{empty}|i,m,0\} = 1 - \rho,$$

$$\Pr\{\text{empty}|\text{empty}\} = 1 - \rho.$$

As shown in **Figure 4**, we have

$$\sum_{k=0}^{m}\sum_{j=0}^{W_k} b_{i,k,j} + b_{\text{empty}} = 1 \tag{3}$$

Moreover, the stationary distribution can be calculated by using the state transition probability.

$$\sum_{k=1}^{m} b_{i,k,0} = \frac{P_{i,\text{fail}}\left(1 - P_{i,\text{fail}}^m\right)}{1 - P_{i,\text{fail}}} b_{i,0,0},$$

$$\sum_{j=0}^{W_0} b_{i,0,j} = \frac{W_0 + 1}{2P_{i,\text{idle}}} b_{i,0,0}, \tag{4}$$

$$b_{\text{empty}} = \frac{1-\rho}{\rho} b_{i,0,0}.$$

From above equations, the $b_{i,0,0}$ can be described as a function of $P_{i,\text{idle}}$, $P_{i,\text{fail}}$, ρ, W_k and T_i, Furthermore, the access probability of every device can be calculated by solving n equations.

$$\tau_i = \sum_{k=0}^{m} b_{i,k,0} = \frac{1 - P_{i,\text{fail}}^{m+1}}{1 - P_{i,\text{fail}}} b_{i,0,0} \tag{5}$$

3.2. System Throughput

The probability in which at least one device is sending a packet is called as transmission probability, P_{tran}.

$$P_{\text{tran}} = 1 - \prod_{j=1}^{n}\left(1 - \tau_j\right) \tag{6}$$

The successful probability of device i means that only device i is transmitting on the medium under condition on the fact that at least one device is transmitting and is represented by $P_{i,\text{suc}}$. In addition, the coordinator can decode the packet correctly.

$$P_{i,\text{suc}} = \frac{\tau_i \prod_{j=1}^{n}\left(1 - \tau_j\right)}{\left(1 - \tau_i\right) P_{\text{tran}}}\left(1 - \text{PER}_i\right) \tag{7}$$

Let $P_{\text{suc}} = \sum_{i=1}^{n} P_{i,\text{SUC}}$ denote the total successful probability of all devices. Once the transmission is successful, the device receives a ACK packet with no payload from the coordinator, whereas the device receives NACK packet or nothing after the timing to receive the ACK packet if the transmitted packet is collided or unable to decode. Consequently, the duration time to transmit a packet successfully, T, is assumed to equal to the duration time of failed transmission, hereafter T is called as the successful transmission time. The successful transmission time is the total duration time to transmit a packet, includes the duration time to transmit a data packet $\left(T_{\text{data}}\right)$, interframe spacing $\left(T_{\text{pSIFS}}\right)$, ACK packet $\left(T_{\text{ACK}}\right)$ and delay time $\left(\alpha\right)$.

$$T = T_{\text{DATA}} + T_{\text{ACK}} + 2T_{\text{pSIFS}} + 2\alpha \tag{8}$$

Let T_P, T_{PHY}, T_{MAC}, T_{BODY} and T_{FCS} denote the duration time to transmit a preamble, PHY header, MAC header, MAC body and FCS, respectively. Therefore, the duration time to transmit a data packet is given by

$$T_{\text{DATA}} = T_P + T_{\text{PHY}} + T_{\text{MAC}} + T_{\text{BODY}} + T_{\text{FCS}},$$

$$= \frac{\text{Preamble} + \text{PHY header}}{R_s} + \frac{8\left(\text{MAC header} + x + \text{MAC footer}\right)}{R_{\text{hdr}}}. \tag{9}$$

Since an immediate ACK/NACK carries no payload, its transmission time is represented as follows.

$$T_{ACK} = T_P + T_{PHY} + T_{MAC} + T_{FCS},$$
$$= \frac{\text{Preamble} + \text{PHY header}}{R_s} + \frac{8(\text{MAC header} + \text{MAC footer})}{R_{hdr}}. \tag{10}$$

Finally, the throughput of device i is described as

$$\text{Thro}_i = \frac{P_{tran}P_{i,\text{suc}}xr}{(1-P_{tran})T_s + P_{tran}P_{suc}T + P_{tran}(1-P_{suc})T}$$
$$= \frac{P_{tran}P_{i,\text{suc}}xr}{(1-P_{tran})T_s + P_{tran}T}, \tag{11}$$

and the system throughput becomes

$$\text{Thro} = \sum_{i=1}^{n} \text{Thro}_i \tag{12}$$

The throughput of scheme 2 is also represented by (12). However, several devices access the channel via the relay and the wireless LAN 1, therefore, the concentration at the wireless LAN 2 is avoided and the successful probability of all devices increases. As a result, the throughput of system is expected to increase.

3.3. Delay

The average access delay D, defined as the time elapsed between the time instant when the frame is put into service and the instant of time the frame terminates a successful delivery. Under the assumption of no retry limits, this computation is straightforward. In fact, we may rely on the well known Little's Result, which states that, for any queueing system, the average number of customers in the system is equal to the average experienced delay multiplied by the average customer departure rate. The application of Little's result to our case yields:

$$D = \frac{x\lambda}{\dfrac{\text{Thro}}{n}} = \frac{x\lambda}{\text{Thro}_i} \tag{13}$$

The delay computation is more elaborate when a frame is discarded after reaching a predetermined maximum number of retries m. In fact, in such a case, a correct delay computation should take into account only the frames successfully delivered at the destination, while should exclude the contribution of frames dropped because of frame retry limit (indeed, the delay experienced by dropped frames would have no practical significance).

To determine the average delay in the finite retry case, we can still start from Little's Result, but we need to replace λ in (13) with the average number of frames that will be successfully delivered. Thus, (13) can be rewritten by

$$D = \frac{x\lambda\beta_i}{\text{Thro}_i} \tag{14}$$

here, β_i denotes the probability that a randomly chosen frame will be successfully transmitted before the retransmission reaches the retry limit. Therefore, the β_i is represented as follows.

$$\beta_i = P_{i,\text{suc}} \sum_{j=0}^{m-1} (1-P_{i,\text{suc}})^j \tag{15}$$

For Scheme 2, the delay due to the multiple access at the wireless LAN 1 and 2 is similar to (14). However, the delay due to the capability of relay also should be considered. The delay due to the relay is calculated by $\dfrac{\displaystyle\sum_{j\in Q}\text{Thro}_j}{C}$, here Q denote the set of devices that access the relay and the C is the capability of relay. Therefore, the average delay of information data that is transmitted via relay is represented as follows.

$$D = \frac{x\lambda\beta_j}{\text{Thro}_j} + \frac{\sum_{j \in Q}\text{Thro}_i}{C} \qquad (16)$$

The delay of scheme 2 is the maximal delay of information data that is transmitted to wireless LAN 1 and 2.

3.4. Bandwidth Efficiency

In order to compare the system with and without relay, the bandwidth efficiency is adopted. The bandwidth efficiency of both schemes 1 and 2 is calculated as the ratio of total throughput of system and the total generated data. Notice that the total throughput of scheme 1 and 2 is different.

$$\delta = \frac{\text{Thro}}{nx\lambda}. \qquad (17)$$

4. Numerical Evaluation

The system model is the same as mentioned above and the parameters in **Table 1** are used. The average distance between all devices and the wireless LAN 1 and 2 is respectively 500 m and 250 m. The relay is set at halfway between the devices and the wireless LAN 1. The delay of propagation is taken into account. The capability of relay is assumed to be 300 Mbps. The noise-free is also assumed. At first, the performance of scheme 1 is illustrated.

The throughput of scheme 1 base on lambda and the number of devices is described in **Figure 5** and **Figure 6**, respectively. The generated data is the total data that is generated at all devices, however the generated data isn't always successfully transmitted due to the collision and the time out. Therefore, the throughput of system is considerably smaller than the generated data, especially when the number of devices and/or the lambda are high. Moreover, the delay of scheme 1 also increase when the number of devices and/or the lambda increase (**Figure 7**). These are the reason the scheme 2 is taken into consideration as description in Section 1.2.

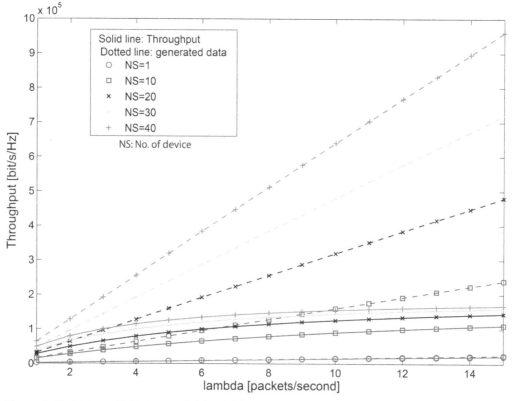

Figure 5. Throughput of scheme 1 based on lambda.

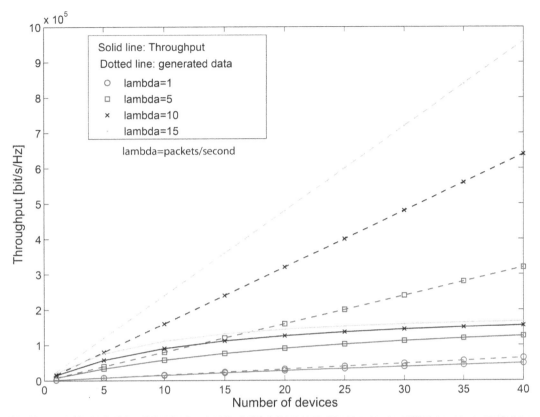

Figure 6. Throughput of scheme 1 based on the number of devices.

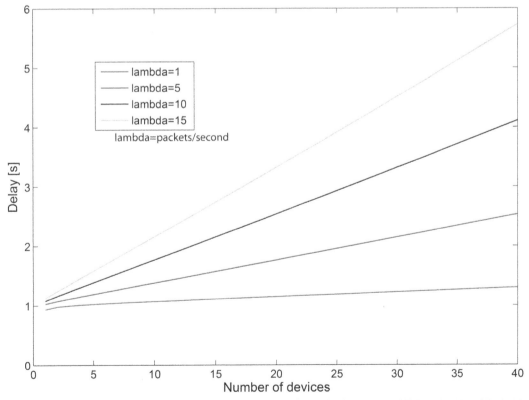

Figure 7. Delay of scheme 1.

The comparison on the delay and the bandwidth efficiency of both schemes 1 and 2 is respectively described in **Figure 8** and **Figure 9**, where the number of devices is fixed to be 10 and 40. For scheme 2, since the concentration at the wireless LAN 2 is avoided, the collision probability decreases. Therefore, the throughput of system increase and then the delay as well as the bandwidth efficiency increase and be higher than that of scheme 1, especially when the number of devices and/or the lambda are large. When the number of devices and the lambda are low, the difference of schemes 1 and 2 is small. Notice that the scheme 2 is more complicated due to the adding of relay and controlling the transmission of devices. It means that there are the trade off between the performance and the complication of scheme 2.

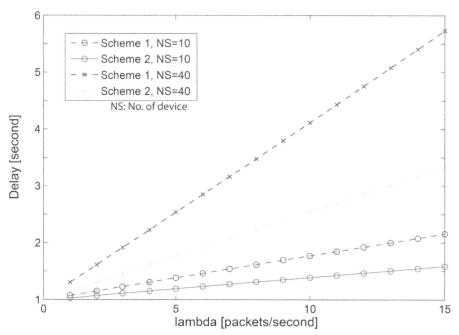

Figure 8. Delay base on lambda.

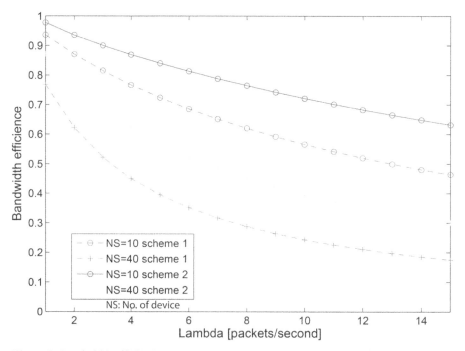

Figure 9. Bandwidth efficiency.

5. Conclusions

The wireless system in a hospital has been taken into consideration and the performance is analyzed based on the standard IEEE802.15.6. The DTMC method was proposed to calculate the access probability and then the collision probability, the successful probability, the throughput, the delay, the bandwidth efficiency of system have been theoretically calculated. The performance of system with and without relay is also numerically compared, the bandwidth efficiency of scheme 2 is higher while the delay is smaller than that of scheme 1 when the number of devices and/or the packet rate are large. However, the scheme 2 is more complicated due to the controlling transmission of devices and the adding of relay.

The devices were assumed to transmit the information data to either the wireless LAN 1 or the wireless LAN 2. However, the control method hasn't been explained clearly. Moreover, the CSMA/CA was adopted, another access protocol [15] wasn't taken into account leave them to our future works.

References

[1]   Wireless Personal Area Network Working Group (2012) IEEE Standard 802.15.6, Wireless Body Area Networks. *IEEE Standards*, 1-271.

[2]   Kwak, K.S., Ullah, S. and Ullah, N. (2010) An Overview of IEEE 802.15.6 Standard. *Proceedings of 3rd International Symposium on Applied Sciences in Biomedical and Communication Technologies*, Rome.

[3]   Martelli, F., Buratti, C. and Verdone, R. (2011) On the Performance of an IEEE 802.15.6 Wireless Body Area Network. *Proceedings of European Wireless* 2011, Vienna.

[4]   Ullah, S., Chen, M. and Kwak, K.S. (2012) Throughput and Delay Analysis of IEEE 802.15.6-Based CSMA/CA Protocol. *Journal of Medical Systems*, 36, 3875-3891. http://dx.doi.org/10.1007/s10916-012-9860-0

[5]   Ullah, S. and Henry Higgin, H., Braem, B., Latre, B., Blondia, C., Moerman, I., Saleem, S., Rahman, Z. and Kwak, K.S. (2012) A Comprehensive Survey of Wireless Body Area Networks on PHY, MAC, and Network Layers Solutions. *Journal of Medical Systems*, 36, 1065-1094. http://dx.doi.org/10.1007/s10916-010-9571-3

[6]   Jung, B.H., Akbar, R.U. and Sung, D.K. (2012) Throughput, Energy Consumption, and Energy Efficiency of IEEE 802.15.6 Body Area Network (BAN) MAC Protocol. *IEEE 23rd International Symposium on Personal Indoor and Mobile Radio Communications (PIMRC)*.

[7]   Rezvani, S. and Ghorashi, A. (2012) A Novel WBAN MAC Protocol With Improved Energy Consumption and Data Rate. *KSII Transactions on Internet and Information System*, 6, 2302-2322.

[8]   Marinkovi, S.J., Popovici, E.M., Spagnol, C., Faul, S. and Marnane, W.P. (2009) Energy-Efficient Low Duty Cycle MAC Protocol for Wireless Body Area Networks. *IEEE Transactions on Information Technology in Biomedicine*, 13, 915-925. http://dx.doi.org/10.1109/TITB.2009.2033591

[9]   Ullah. S., An, X. and Kwak, K. (2009) Towards Power Efficient MAC Protocol for In-Body and On-Body Device Networks. *Agent and Multi-Agent System: Technologies and Application*, 5559, 335-345.

[10]   Zhen, B., Li, H.B. and Kohno, R. (2008) IEEE Body Area Networks and Medical Implant Communications. *Proceedings of the ICST 3rd International Conference on Body Area Networks*, Tempe.

[11]   Rashwand, S., Misic, J. and Khazaei, H. (2011) IEEE 802.15.6 under Saturation: Some Problems to be Expected. *Journal of Communications and Networks*, 13, 142-148. http://dx.doi.org/10.1109/JCN.2011.6157413

[12]   Rashwand, S. and Misic, J. (2011) Performance Evaluation of IEEE 802.15.6 under Non-Saturation Condition. *Proceedings of IEEE Global Telecommunications Conference (GLOBECOM)*, Kathmandu.

[13]   Li, C.L., Geng, X., Yuan, J. and Sun, T. (2013) Performance Analysis of IEEE 802.15.6 MAC Protocol in Beacon Mode with Superframes. *KSII Transaction on Internet and Information Systems*, 7, 1108-1130. http://dx.doi.org/10.3837/tiis.2013.05.010

[14]   Rashwand, S. and Misic, J. (2012) Effects of Access Phases Lengths on Performance of IEEE 802.15.6 CSMA/CA. *Journal of Computer Networks*, 56, 2832-2846. http://dx.doi.org/10.1016/j.comnet.2012.04.023

[15]   Bianchi, G. (2000) Performance Analysis of the IEEE 802.11 Distributed Coordination Function. *IEEE Journal on Selected Areas in Communications*, 18, 535-547. http://dx.doi.org/10.1109/49.840210

2

Hard Decision-Based PWM for MIMO-OFDM Radar

Omar Daoud

Communications and Electronics Engineering Department, Philadelphia University, Amman, Jordan
Email: odaoud@philadelphia.edu.jo

Abstract

For the purpose of target localization, Multiple Input Multiple Output-Orthogonal Frequency Division Multiplexing (MIMO-OFDM) radar has been proposed. OFDM technique has been adopted in order to a simultaneous transmission and reception of a set of multiple narrowband orthogonal signals at orthogonal frequencies. Although multi-carrier systems such as OFDM support high data rate applications, they do not only require linear amplification but also they complicate the power amplifiers design and increase power consumption. This is because of high peak-to-average power ratio (PAPR). In this work, a new proposition has been made based on the Pulse Width Modulation (PWM) to enhance the MIMO-OFDM radar systems' performance. In order to check the proposed systems performance and its validity, a numerical analysis and a MATLAB simulation have been conducted. Nevertheless of the system characteristics and under same bandwidth occupancy and system's specifications, the simulation results show that this work can reduce the PAPR values clearly and show capable results over the ones in the literature.

Keywords

Multiple Input Multiple Output, Orthogonal Frequency Division Multiplexing, RADAR, Peak to Average Power Ratio, Pulse Width Modulation

1. Introduction

Many researchers have turned their attentions toward the Orthogonal Frequency Division Multiplexing (OFDM) scheme in order to provide high data rate applications under maintaining the spectral efficiency. Therefore, its clearly deployed in many broadband communication systems and protocols such as WiFi, WiMax, 4G and advanced LTE, Bluetooth-2. However, due to a high Peak-to-Average Power Ratio (PAPR), linear amplifiers suffer from low power efficiency under the utilization with multicarrier systems [1]-[5]. As a consequence, the cost of such devices; power amplifiers, mixers and analogue to digital convertor will be increased [1] [6]. OFDM signals are easily generated and produced by applying the Fast Fourier Transform (FFT) processing block and its Inverse (IFFT) [7]. This is due to its high speed processing in performing the needed operations, such as the transformation process, filtering and correlation [8]. On the other hand, Multiple Input Multiple Output (MIMO) concept has been found in the literature to enhance either the transmission capacity or the link robustness (independent or dependent information streams are transmitted via parallel sub-channels simultaneously). In contrast, in MIMO radar there is no need to construct parallel subchannels which are fully dependent on a multipath environment. This is due to that all transmitted information is known at the receiver side. Thus, the channel matrix is used only for the purpose of sensing the environment, as an example to determine the number of the targets, their locations and velocities [9]-[12].

In order to enhance the MIMO radar, which is adopting the OFDM technique, a new work has been proposed in this paper based on hard decision-based PWM technique to tackle one of the main deficiencies found in OFDM; namely PAPR.

This deficiency appears due to the addition process with different frequencies and phases of numerous waves, which leads to the need of high dynamic ranges transmitters. The predicted PAPR values in OFDM signal can be formed as [7]:

$$\text{PAPR} = \frac{\max_{0 \le t \le T}\left(\left|\mathbb{Z}(t)\right|^2\right)}{\left(\frac{1}{T}\int_0^T\left(\left|\mathbb{Z}(t)\right|^2\right)\mathrm{d}t\right)} \tag{1}$$

Here, $\mathbb{Z}(t)$ is the transmitted OFDM symbol and could be found as $\mathbb{Z}(t) = \left(\sqrt{\frac{\left(\sum_{k=0}^{N-1}\left(X_k \cdot \mathrm{e}^{(j2\pi f_k t)}\right)\right)^2}{N}}\right)$,

which results from the modulation process of an N symbols data block; X_k with f_k, which are a set of orthogonal subcarriers for $k = 0, \cdots, N-1$. The duration of OFDM symbol is denoted by T, which used to maintain the orthogonality for all values of t less than or equal to T. Moreover, to maintain the total transmitted power, the term $\left(1/\sqrt{N}\right)$ has been imposed.

Such deficiency causes transmission amplification and other circuitry limitations. Therefore, to overcome this problem, average signal power must be kept low to allow the transmission process of the higher average power to be in a fixed level. Then an improvement of the reception process will be attained based on improving the signal to noise ratio.

There are several propositions and techniques that either tackle the PAPR effects or address the linearity and power efficiency issues, such as filtering and clipping techniques; coding based techniques; artificial intelligence based techniques; and signal representation techniques as the envelope elimination and restoration techniques and the phase shifted sequence ones. This is in order to optimize a solution at the expense of several challenges, such as the degradation of the Bit Error Rate (BER); the decrement of the spectral efficiency due to the side information (SI) transmission; and the computational complexity [13]-[16].

This paper addresses the proposition of a new technique based on using the pulse width modulation (PWM) to overcome the PAPR problem effect. Consequently, the overall performance will be enhanced for the MIMO radar, which adopting the OFDM technique. As a result of considering the use of PWM, a basis of controlling the power electronics [17], an optimum solution will be provided to optimize the vital parameters of the existing work such as the speed, and the area. Moreover, the proposed work has been compared with either conventional OFDM systems in the literature or our previous published work in order to show the performance improvements before applying it to a MIMO-OFDM radar system.

PWM signal is easily generated by comparing the reference signal with a carrier one. Mainly, the input signal is used to determine the width of the generated PWM signal. This is clearly shown in the following mathematical representation

$$\text{PWM}(t) = \text{sgn}\left(r(t) - c(t)\right) \tag{2}$$

where the generated PWM signal depends on the sign function of the subtraction process between the compared reference signal; $r(t)$ and the carrier signal; $c(t)$.

As basic PWM signal generation, there are two methods that help in producing the variable pulses widths; direct digital generation and uniformly sampled PWM. They can be distinguished by the focusing on the controlling criteria. In this paper, the second technique will be chosen, where a triangle clock signal is used to generate the uniformly sampled PWM signal. This is due to that it does not need high frequency clock signal. Moreover, the triangle clock signal is chosen over the other two types, Sawtooth or the inverted Sawtooth, due to that it has low number of dominant higher harmonics. The achieved benefit here concluded in reducing the needed system bandwidth [17] [18].

The rest of this work is introduced as follows: Section 2 describes the model of MIMO-OFDM radar signals based on PWM along with the analytical formulation in addition to the computational complexity. Section 3 presents simulation results and hardware implementations; finally, the conclusion is represented in Section 4.

2. MIMO-OFDM Radar Signal Model-Based PWM

2.1. MIMO-OFDM Radar Systems Structure

In [1], OFDM technique has the advantage of combating the frequency selective fading drawback for a narrow-band system. This turns the researchers toward making use of such advantage to be imposed in MIMO radar systems. Therefore, the radar system performance will be improved by performing the target localization separately. This will be attained by making use of the combination of different orthogonal and narrowband sub-signals. As a result, the frequency selective fading deficiency is overcome by the frequency diversity utilization. Moreover, the complexity of MIMO-OFDM radar transmitters will be maintained at low level, since the used sub-bad waveforms designed to have same characteristics as the narrowband MIMO radar ones.

The baseband MIMO-OFDM transmitted matrix is defined in (1) by $\mathbb{Z}(t)$, where an N sets of orthonormal signals have been sent simultaneously. The digital-to-analogue convertor (DAC) has been used to efficiently generate each OFDM symbol from $\mathbb{Z}(t)$ after the IFFT stage. For practical implementation using the IFFT, $\mathbb{Z}(t)$ should be oversampled [1] as shown below:

$$\left[\mathbb{Z}_{i,j}(1)\cdots\mathbb{Z}_{i,j}(L)\right] = \mathrm{IFFT}\left(\left[\boldsymbol{X}_{i,j}(0)\cdots\boldsymbol{X}_{i,j}(k-1)\mathbf{0}_{1\times(L-k)}\right]\right) \tag{3}$$

Here, the IFFT of $\mathbb{Z}(t)$ matrix of the size of $(i \times j)$ will have an L samples and defining its l-th sampling by $\mathbb{Z}_{i,j}(l)$, the element of i-th row and j-th column of the matrix $\boldsymbol{X}(k)$ is given by $\boldsymbol{X}_{i,j}(k)$, and $\mathbf{0}_{1\times(L-k)}$ for zero padding when $L \geq k$. Accordingly, the produced i-th OFDM symbol will be processed and transmitted from the i-th antenna.

The next step after generating the OFDM baseband waveforms is the imposing of a guard interval process at the beginning of each OFDM symbol. This is attained by attaching a copy of the later OFDM symbol part at the beginning; namely a cyclic prefix process, which is introduced in order to maintain the orthogonality between the used sub-bands by MIMO operation and will be accomplished by making use of windowing techniques. Moreover, the MIMO-OFDM radar will make use of it in order to compensate for the shifts in time. This is clearly shown under the case of multiple targets at different ranges, where it guarantees the existence of the needed phase delay information inside the used window. This is true under a predefined separation range, which is based on the antenna array dimension and the used lengths for the transmitted symbol period and the cyclic prefix length. Therefore, the final transmitted baseband matrix is given as

$$\mathbb{Q}(t) = \mathbb{Z}(t-T_p)W\left(\frac{t-T_p}{T_s}\right) + \mathbb{Z}(t+T_s-T_p)W\left(\frac{t}{T_p}\right) \tag{4}$$

Here, T_p is the length of the cyclic prefix, T_s is the $\mathbb{Z}(t)$ elements period and equals $1/f_k$, W is a window function that has a value of unity when $t \in [0,1]$.

Moreover, in order to maintain the orthogonality condition, the antenna elements displacement should satisfy curtain threshold. In this work, to detect and estimate a target within $180°$, the displacement; D should satisfy

$$D \leq \left(c \times \left(T_p - T_{\mathrm{target}}\right)\right) \tag{5}$$

where, c is the speed of light and T_{target} is the target impulse response length.

The imposing process of the cyclic prefix is clearly described in **Figure 1**, where a simultaneous transmission of three OFDM symbols from three antennas is accomplished.

In this work, and after the imposing of the guard interval; *i.e.* the cyclic prefix, a new processing block has been inserted to analyze the PAPR performance. This is to free the channel from the inter symbol interference (ISI) drawback. Moreover, this choice will reduce the hardware area under the consideration of hardware implementation.

2.2. Proposition of PWM

Returning to the implementation of the baseband OFDM signal; *i.e.* $\mathbb{Z}(t)$; in (1), and the possibility of producing high PAPR values. The overall system performance will be enhanced if the PAPR effect is efficiently reduced. This work focusing on proposing PWM based work that will produce constant amplitude signals by making use of the consecutive samples slope. It's considered as a novel technique that is attaining the maximum power amplifier efficiency issues, which will permit the ability of using nonlinear devices easily. Complementary cumulative distribution function (CCDF) curves give the statistical characteristics of the PAPR distribution in OFDM systems [19] [20]. CCDF curves show the probability of exceeding the PAPR a certain threshold for different signal to noise ratio values, *i.e.* the maximum power amplifier efficiency will be attained at the minimum CCDF value.

The proposed PWM work starts with reshaping the signal, $\mathbb{Z}(t)$, in (1) as blocks, each with length equals to the IFFT points as follows

$$\mathbb{Z}(n,m) = \left(\sum_{k=0}^{N(m)-1} \left[X(k,m) \right] \left[e^{\left(j\frac{2\pi kn}{N(m)} \right)} \right] \right) \bigg/ \left(\sqrt{N(m)} \right) \tag{6}$$

Here, m stands for the block index as defined in (7), and $N(m)$ denotes the block length.

$$m = \begin{cases} 0, & n < N(0) \\ \rho, & \sum_{i=0}^{\rho-1} N(i) \le n < \sum_{i=0}^{\rho} N(i), \rho \ge 1 \end{cases} \tag{7}$$

The reshaped result from (6) has been used to be processed in the production of a constant amplitude signal based on the PWM technique. This is clearly depicted and shown in **Figure 2**.

As shown in **Figure 2**, the conversion process is divided into the following stages:

First stage:
- In order to distinguish each symbol after the conversion process an extra zero sample has been added at the beginning of each OFDM symbol.
- The sampling rate has been increased in order to enhance the accuracy of the conversion process; $\mathbb{Z}(n,m)$ will have extra samples under the use of new sampling rate $\dot{N}(m)$. This is clearly shown in (8) where $\mathbb{Z}(n,m)$ has been converted into a three vectors-matrix.

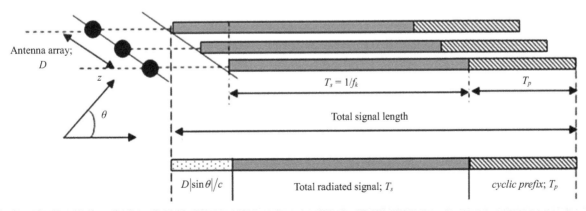

Figure 1. 3-MIMO-OFDM symbols transmission process.

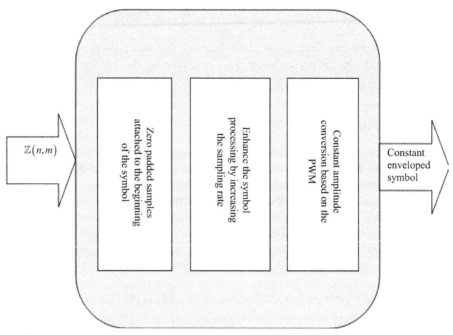

Figure 2. The conversion procedure.

$$
\mathbb{Z}(3,i) = \begin{cases} \mathbb{Z}(n,m) \\ \mathbb{Z}(n,m) - \dfrac{1}{1+\grave{N}(m)}, \\ \mathbb{Z}(n,m) + \dfrac{1}{1+\grave{N}(m)} \end{cases} \tag{8}
$$

where i stands for the sample value; $i \in \left[1, \grave{N}(m) + N(m)\right]$.

Second Stage:

In this stage, the oversampled version of the OFDM symbol will be processed in order to produce a constant envelope version. For simplicity, the slope between two consecutive samples has been chosen as a comparison criterion. This criterion is depicted as follows in (9).

$$
\mathbb{Z}(i) = \begin{cases} \mathbb{Z}(i) = \mathbb{Z}(1,(i-1)), & \text{then } \mathbb{Z}(i) = 0 \\ \mathbb{Z}(i) < \mathbb{Z}(2,(i-1)), & \text{then } \mathbb{Z}(i) = -\text{slope} \\ \mathbb{Z}(i) > \mathbb{Z}(3,(i-1)), & \text{then } \mathbb{Z}(i) = +\text{slope} \end{cases} \tag{9}
$$

These two stages are clearly described in **Figure 3**. It's divided into two main parts denoting the stages consequently.

As described earlier, **Figure 3(a)** depicts the process of the first stage and **Figure 3(b)** represents the procedure of the second stage. In **Figure 3(b)**, the pre-process is divided into two main parts; the one that is responsible for simplifying the distinguishing process by adding a pre-known sample(s) at the beginning of the conventional symbol. In this work, the zero sample will do the expected results in either the transmission part or the reception one.

This is clearly found in **Figure 4**, where a zero sample has been attached at the beginning of the reshaped signal, *i.e.* $\mathbb{Z}(n,m)$. The second part deals with sampling rate of the modified $\mathbb{Z}(n,m)$; here extra samples have been imposed between the two consecutive samples by enhancing the sampling rate to be $\grave{N}(m)$. **Figure 5**, shows the enhancement sampling rate for the signal with 108 samples instead of 12 samples.

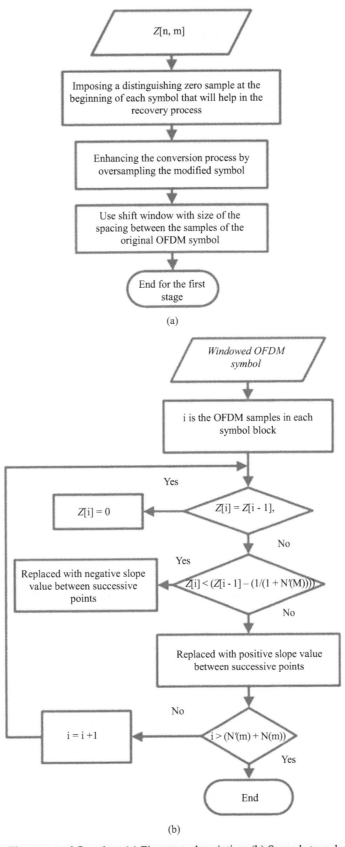

Figure 3. The proposed flowchart (a) First stage description; (b) Second stage description.

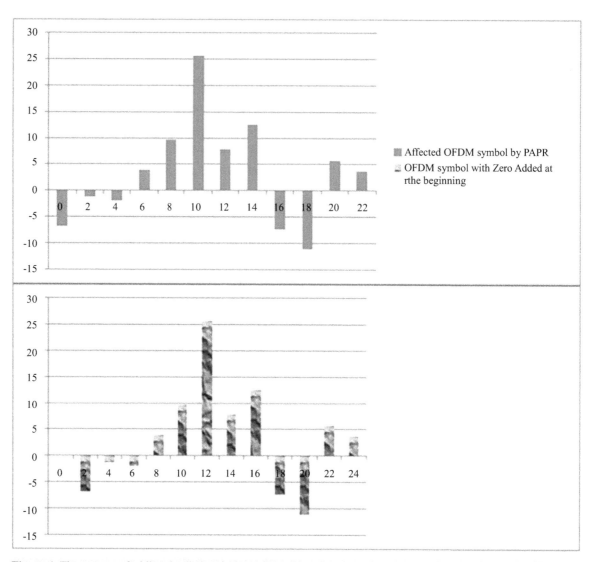

Figure 4. The process of adding the distinguishing zero sample.

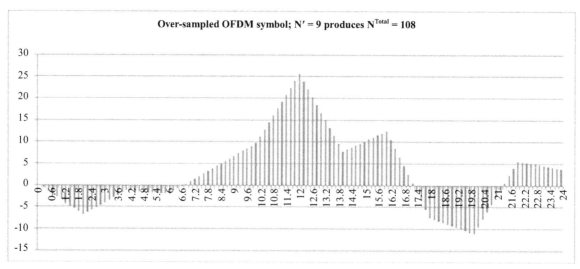

Figure 5. The over-sampled OFDM symbol.

Figure 6 depicts the output results that are achieved from the second stage in the proposed work and shown in **Figure 3(b)**, which is drawn from the proposed comparison formula in (9). In this figure, the idea of the proposed work has been showed up; *i.e.* the OFDM symbol has been converted to a constant enveloped symbol based on the slope between the adjacent samples.

The variation has been reduced and the peaks values have been diminished, consequently the PAPR values will be reduced. In addition, to simplify the proposed work, the hermitian structure of the OFDM systems could be exploited [21]. At this stage, the OFDM symbol is ready to be transferred to the next block; MIMO block to process the rectified symbol based on the Vertical-Bell Laboratories Layered Space-Time (V-BLAST) criterion. V-BLAS technology is used to attain the system capacity/throughput enhancement, which is expressed in terms of bits/symbol.

The transmitter stages are shown in **Figure 7**. The transmitted signal through I antennas will be guaranteed to be with the minimum PAPR value, since the resultant unaffected MIMO-OFDM signal will be based on the constant enveloped transformed signal using PWM. As a result, the novelty of this work rises from the way of dealing with the OFDM symbol for such application. This is in addition to the way of how to reduce the CCDF curves values that are guaranteed to remains at their minimum levels.

In the receiver side, the signal modelling will be determined based on the sent and received signals between/ among the transmitter and the object(s), which will help in determining the objects specifications. The issues of determining the objects directions and locations are considered out of scope of this work and will be discussed in another work. Thus, it is focusing on how to overcome the rise problem due to the use of the FFT and its inverse in modelling the OFDM system.

After the transmission through a channel from different transmitting antennas, in the receiver side, the main task of the receiver is to recover the original OFDM signal from the modified one. Accordingly, the used recovering

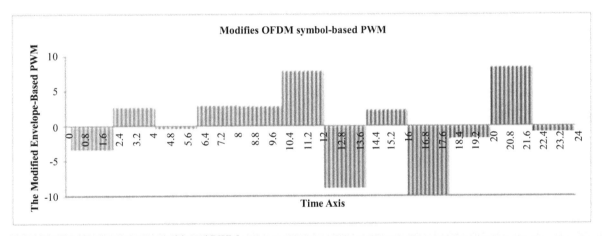

Figure 6. Modified OFDM symbol based PWM.

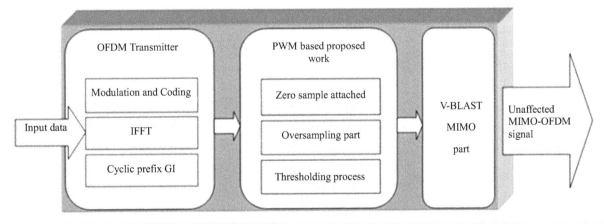

Figure 7. The proposed MIMO-OFDM radar transmitter.

procedure will be divided into two main stages; firstly, proposing an algorithm to recover the OFDM symbol from the constant enveloped received symbols, and secondly, a signal processing stage based on removing the extra imposed samples. This procedure is clearly shown and described in **Figure 8**. It contains two parts depicted the proposed processing stages; **Figure 8(a)** and **Figure 8(b)** according to stage 1 and stage 2 consequently.

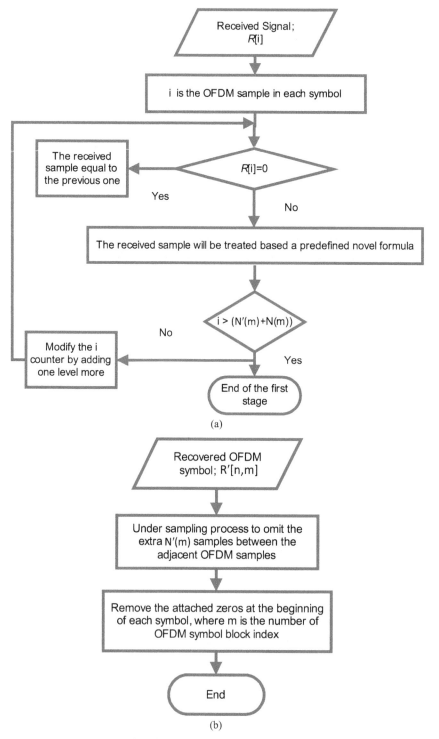

Figure 8. Reception stage flowchart (a) First stage procedure; (b) Second stage procedure.

The reproduction process of regenerating OFDM signal from the received is clearly depicted in **Figure 8**. It is divided into two main parts; the reconversion stage and the de-processing stage; shown in **Figure 8(a)** and **Figure 8(b)** respectively.

The given procedure in **Figure 8(a)** starts with checking the received samples to clarify whether it is the beginning of new symbol or not as shown in (10). Moreover, a novel formula has been proposed to reproduce the generation of the original OFDM symbols based on three different variables; the received sample, the difference in time between a consecutive samples and the previous sample.

$$R[i] = \begin{cases} 0 \Rightarrow \left\{ \text{if } R[i-1] \begin{cases} \text{existed} \\ \text{does not exist} \end{cases} \Rightarrow \begin{cases} R[i] = R[i-1] \\ \text{new symbol starts} \end{cases} \right. \\ \text{O.W.} \begin{cases} \text{the sample will be treated based on a novel formula,} \\ \text{which is depending on the received sample and the time} \\ \text{interval and the previous received sample} \end{cases} \end{cases} \tag{10}$$

Furthermore, in **Figure 8(b)** the conversion process has been completed by fulfilling the removal of the extra added samples. The novelty in this work has been shown in how to deal with the OFDM symbol; imposing hard decision criterion based on the PWM instead of sending the original OFDM signals.

The next section describes the results from the proposed work simulation against the conventional techniques; they are based on both of CCDF and SER curves. These two criteria are used to validate the OFDM systems performance, where the lower the values the higher the system performance.

3. Tested System Performance and Simulated Results

The proposed MIMO-OFDM radar system that has been described in **Figure 7** is used to localise a composite target based on a four-element array. The composite target has been placed far away from the array, which contains three dielectric elements. They are differed in physical dimensions that are given in terms of the carrier frequency but have same dielectric permittivity. Moreover, the used array is assumed to contain isotropic elements with equal spacing. The scope of this work covers the system performance based on tackling the PAPR problem in the MIMO-OFDM radar. This leaves a room to enhance the performance from different point of views for future work.

At this stage, the performed MATLAB simulation has the following specifications:
- Carrier frequency of 3 GHz,
- Extra 9 samples have been added between the consecutive samples; $N'(m) = 9$,
- IFFT length of 1024 point,
- $T_p = 0.25 \times T_s$,
- Carrier spacing = (1/32) GHz,
- 64-QAM modulation technique,
- Vertical-Bell Laboratories Layered Space-Time (V-BLAST) MIMO system will be used for the four-element array transmission. This is in order to boost the system performance in terms of bits/symbol.

To imitate a real scenario, the shown proposed work in **Figure 7** has been imposed twice; one for the real OFDM symbol part and the other one for the imaginary part. Moreover and in order to test the system's performance, the results have been divided into two parts; the complementary cumulative distribution function (CCDF) part and the sample error rate (SER) part. **Figure 9** will depict the value of proposed work SER and how promising the achieved values are, while **Figure 10** will clarify the enhancement in combating the effect of the PAPR from the CCDF point of view, which is considered as a performance metric independent of the transmitter amplifier. It is defined by how often the PAPR is higher than a given threshold; $PAPR_o$. It is expressed by:

$$CCDF(PAPR_o) = Probability\{PAPR > PAPR_o\} \tag{11}$$

As depicted in **Figure 9**, the proposed PWM enhances the recovery process of the original OFDM samples. This work is divided into two experiments; the first one based on using the previous sample in order to predict the received sample, while the other one is based on using the average of all samples within the window. The one that is based on the average gives a better SER under the optimization issue between the SER and the delay

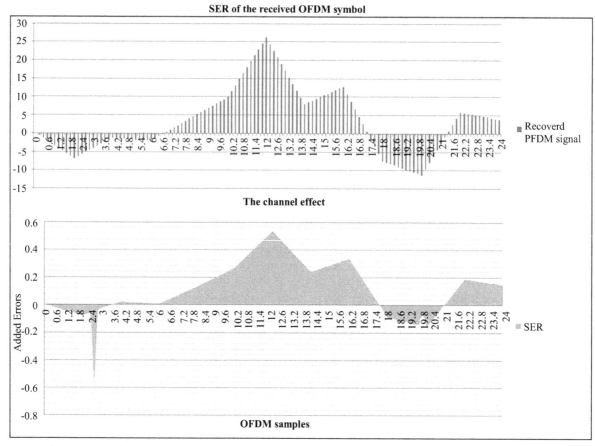

Figure 9. The SER and the recovered OFDM samples.

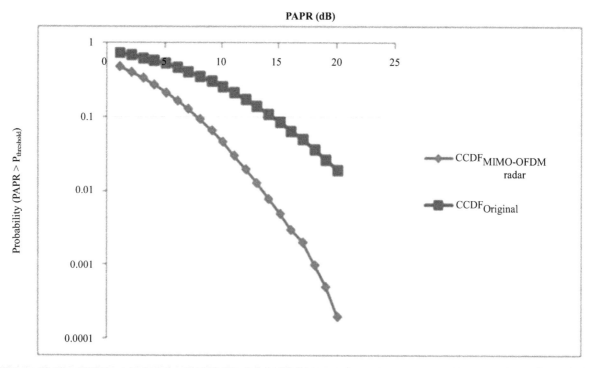

Figure 10. The CCDF results based on 64QAM modulation technique.

Table 1. The Simulation results of the proposed technique based on PWM to the literature work.

	CCDF (2%)		Additional reduction (%)		
Modulation technique	PAPR without coding (dB)	MIMO-OFDM radar based PWM (dB)	Clipping	SLM	PTS
64 QAM	20	11.5	72.1	44.5	11

time. This modification enhances the SER from 9.3×10^{-4} to 8.7×10^{-4}.

Furthermore, the second part of the systems performance checking is shown in **Figure 10** based on the CCDF curves. From the depicted results in **Figure 10**, the proposed work gives a noticeable improvement in the CCDF curves. Using the PWM as a processing technique to enhance the OFDM signal's envelope reduces the CCDF (20 dB) from 19×10^{-3} to 2×10^{-4}. Furthermore, a comparison has been made between our proposed work and the work that found in the literature. This comparison is shown in **Table 1**.

The proposed system performance improvement has bean clearly shown in **Table 1**. It shows that our work has 2% of the PAPR that is exceeds 11.5 dB, where for the same probability percentage the conventional OFDM systems has a PAPR over 20 dB. Moreover and comparing to the literature, the proposed work has an extra reduction percentage between 11% and 72%.

MIMO-OFDM radar work is different from the conventional ones in the literature, where this work has built the comparison stage making use of the slope between the two consecutive samples. Additionally, this work enhances the use of the PWM techniques, where the conventional PWM links the comparison performance to the inserted number of extra samples. In the MIMO-OFDM radar, the performance has been improved without overloading the systems with extra samples since the comparison stage has been linked to the slope between the samples.

4. Conclusions

This work takes high PAPR effect into consideration when proposing the OFDM technique to the conventional MIMO radar systems. High PAPR values could reduce the system performance especially when using nonlinear devices. A new work has been proposed to overcome this deficiency making use of the conventional PWM with some modifications.

A MATLAB simulation has been conducted to validate the analytical model of the proposed work. It consists of two parts: one to check the sample error rate (SER) after the recovery process, while the other one to describe the probability of the PAPR to exceed a certain threshold. Moreover, a comparison with the literature has been made in order to confirm the expected performance modification.

Under same environmental conditions and system specifications, a SER of 8.7×10^{-4} has been achieved compared with the transmission of conventional OFDM signals. This is in addition to enhancing the probability of the PAPR that exceeds 20 dB from 2.1×10^{-2} to 1.7×10^{-4}. The work validity has been checked based on a comparison with the ones in the literature, such as PTS, SLM or Clipping techniques, the proposed work gives an additional PAPR reduction percentage between 11% and 72% over the achieved 11.5 dB value. As a consequence, the transmission throughput will improve.

References

[1] Nee, R. and Prasad, R. (2000) OFDM for Wireless Multimedia Communications. Artech House, Norwood.

[2] Umali, E., Toyama, Y. and Yamao, Y. (2008) Power Spectral Analysis of the Envelope Pulse-Width Modulation (EPWM) Transmitter for High Efficiency Amplification of OFDM Signals. *IEEE Vehicular Technology Conference* (*VTC*), Singapore, 1261-1265.

[3] 3GPP, Tech. Specif. Group Services and System Aspects Service Requirements for Evolution of the 3GPP System (Rel. 8), 3GPP TS 22.278.

[4] Dahlman, E., *et al.* (2008) 3G Evolution: HSPA and LTE for Mobile Broadband. 2nd Edition, Academic Press.

[5] Abeta, S. (2010) Toward LTE Commercial Launch and Future Plan for LTE Enhancements (LTE-Advanced). *IEEE International Conference on Communication Systems* (*ICCS*) *Proceedings*, Singapore, 146-150.

[6] Andrews, J., Ghosh, A. and Muhamed, R. (2007) Fundamentals of WiMAX: Understanding Broadband Wireless Networking. Prentice Hall.

[7]   Jiang, T. and Yu, Y. (2008) An Overview: Peak-to-Average Power Ratio Reduction Techniques for OFDM Signals. *IEEE Transaction on Broadcasting*, **54**, 257-268. http://dx.doi.org/10.1109/TBC.2008.915770

[8]   Saeed, A., Elbably, M. and Abdelfadeel, G. (2009) Efficient FPGA Implementation of FFT/IFFT Processor. *International Journal of Circuits, Systems and Signal Processing*, **3**, 103-110.

[9]   Foschini, G. and Gans, M. (1998) On Limits of Wireless Communications in a Fading Environment When Using Multiple Antennas. *Wireless Personal Communications*, **6**, 311-335. http://dx.doi.org/10.1023/A:1008889222784

[10] Telatar, E. (1995) Capacity of Multi-Antenna Gaussian Channels'. AT & T Bell Laboratories.

[11] Zelst, V. and Schenk, T. (2004) Implementation of a MIMO OFDM-Based Wireless LAN System. *IEEE Transactions on Signal Processing*, **52**, 483-494. http://dx.doi.org/10.1109/TSP.2003.820989

[12] Bekkemani, I. and Tabrkian, J. (2006) Target Detection and Localization Using MIMO Radars and Sonars. *IEEE Transactions on Signal Processing*, **54**, 3873-3883. http://dx.doi.org/10.1109/TSP.2006.879267

[13] Wang, Y. and Luo, Z. (2011) Optimized Iterative Clipping and Filtering for PAPR Reduction of OFDM Signals. *IEEE Transactions on Communications*, **59**, 33-37. http://dx.doi.org/10.1109/TCOMM.2010.102910.090040

[14] Chen, J.C. (2010) Partial Transmit Sequence for PAPR Reduction of OFDM Signals with Stochastic Optimization Techniques. *IEEE Transactions on Consumer Electronics*, **56**, 1229-1234. http://dx.doi.org/10.1109/TCE.2010.5606251

[15] Sohn, I. (2014) A Low Complexity PAPR Reduction Scheme for OFDM Systems via Neural Networks. *IEEE Communications Letters*, **18**, 225-228. http://dx.doi.org/10.1109/LCOMM.2013.123113.131888

[16] Wang, F.P., Kimball, D., Popp, J., Yang, A., Lie, D., Asbeck, P. and Larson, L. (2005) Wideband Envelope Elimination and Restoration Power Amplifier with High Efficiency Wideband Envelope Amplifier for WLAN 802.11g Applications. *IEEE Microwave Symposium Digest*, Long Beach, 12-17 June 2005, 645-648.

[17] Vasca, F. and Lannelli, L. (2012) Dynamics and Control of Switched Electronic Systems: Advanced Perspectives for Modelling, Simulation and Control of Power Converters. Springer Publisher, Berlin.

[18] Koyuncu, M., van den Bos, C. and Serdijn, W. (2000) A PWM Modulator for Wireless Infrared Communication. *Proceedings of the ProRISC/IEEE Workshop on Semiconductors, Circuits, Systems and Signal Processing*, Veldhoven, 30 November-1 December 2000, 351-353.

[19] Ochiai, H. and Imai, H. (2001) On the Distribution of the Peak-to-Average Power Ratio in OFDM Signals. *IEEE Transactions on Communications*, **49**, 282-289. http://dx.doi.org/10.1109/26.905885

[20] Wei, S., Goeckel, D. and Kelly, P. (2010) Convergence of the Complex Envelope of Bandlimited OFDM Signals. *IEEE Transactions on Information Theory*, **56**, 4893-4904. http://dx.doi.org/10.1109/TIT.2010.2059550

[21] Lin, H. and Siohan, P. (2008) OFDM/OQAM with Hermitian Symmetry: Design and Performance for Baseband Communication. *IEEE International Conference on Communications* (*ICC*' 08), Beijing, 19-23 May 2008, 652-656.

3

High-Level Portable Programming Language for Optimized Memory Use of Network Processors

Yasusi Kanada

Central Research Laboratory, Hitachi, Ltd., Yokohama, Japan
Email: Yasusi.Kanada.yq@hitachi.com

Abstract

Network processors (NPs) are widely used for programmable and high-performance networks; however, the programs for NPs are less portable, the number of NP program developers is small, and the development cost is high. To solve these problems, this paper proposes an open, high-level, and portable programming language called "Phonepl", which is independent from vendor-specific proprietary hardware and software but can be translated into an NP program with high performance especially in the memory use. A common NP hardware feature is that a whole packet is stored in DRAM, but the header is cached in SRAM. Phonepl has a hardware-independent abstraction of this feature so that it allows programmers mostly unconscious of this hardware feature. To implement the abstraction, four representations of packet data type that cover all the packet operations (including substring, concatenation, input, and output) are introduced. Phonepl have been implemented on Octeon NPs used in plug-ins for a network-virtualization environment called the VNode Infrastructure, and several packet-handling programs were evaluated. As for the evaluation result, the conversion throughput is close to the wire rate, *i.e.*, 10 Gbps, and no packet loss (by cache miss) occurs when the packet size is 256 bytes or larger.

Keywords

Network Processors, Portability, High-Level Language, Hardware Independence, Memory Usage, DRAM, SRAM, Network Virtualization

1. Introduction

To enable programmability for networking and in-network processing, especially for new network-layer programming for clean-slate virtual-networks [1], network processors (NPs) have been used [2] and will be more widely used in the near future. NPs, which were developed for software-based high-performance networking solutions, make it possible to quickly develop arbitrary protocol and functions in the case of hardware-based solutions as well.

However, there are three problems that make using NPs for such functions difficult. The first problem is lack of portability. Because low-level languages that are similar to assembly languages must be used for developing NP programs, the programs are not portable. Although extended versions of C can usually be used for developing NP programs, essential libraries depend on vendor-specific proprietary hardware and software, and proprietary rights on NP programs are protected by non-disclosure agreements (NDAs) preventing programs and documents concerning an NP being ported. The second problem is high development cost and that the availability of NP program developers is limited. NP program developments require special skills, and the knowledge they require is not widely available; thus, only a limited number of developers have the ability to develop NP programs. In addition, vendor-specific information is required in NP-program development. Consequently, the learning curve of NP-program development is very gentle, the development takes a very long time, and its cost is very high. The third problem is restriction on publishing developed programs, papers, and documents concerning an NP. This is a serious problem for network researchers.

The three above-described problems can be solved by successfully designing and implementing a high-level-language, which can translate programs into NP machine code or a vendor-dependent C program. Programs written in this language must be translated into NP-dependent object programs; however, to solve the problems, the language must be hardware- and vendor-independent.

An important common NP feature concerning high-performance packet processing (to avoid packet drops caused by cache misses) is to use static random-access memory (SRAM) and dynamic random-access memory (DRAM) by different methods with explicit awareness by the programmer, making programming difficult and time-consuming. Although memory allocation is not the only issue that causes the above three problems, this is the most important and serious issue because NPs are optimized for wire-rate processing and memory abuse immediately prevents it and severely reduces the performance. In particular, whole packets are stored in DRAM, and only the headers, which must be modified, removed, or added, are cached in SRAM because if data stored in DRAM is accessed by a CPU core, access takes an excessively long time, and wire-rate processing is impossible. The rest of the packets are just forwarded to the next network node without modification in the NP. This is common because it is necessary for NPs to store packets in memory while processing them, but the size of SRAM is limited, so whole packets cannot be stored in short-access-time memory, i.e., SRAM.

When programming a packet-processing program for NPs, programmers must use an assembly language or C with assembly-level features, and must be very careful to get high performance. When using general-purpose CPUs, programmers can use high-level language and do not have to distinguish SRAM (or cache) and DRAM because they are automatically selected when programs load and store data. However, NP programmers must usually know whether the packet to be processed is on SRAM or DRAM (or both) because this knowledge is critical for attaining stable (i.e., mishit-less) wire-rate processing. Two types of NP architectures are available. In one of them, such as Intel IXP, the SRAM and the DRAM are different classes of memory with different addresses. In the other type, such as Cavium Octeon®, the SRAM can be accessed as cache or registers, in a similar manner to general-purpose CPUs, but programmers must still be aware of the SRAM/DRAM distinction because the NP handles them in different ways. These cases are explained in more detail in Section 2.

Although it is a promising approach to design a new open and portable high-level language and to implement a high-performance language processor, i.e., a compiler and run-time routines, it is still very hard to solve the above three problems because of the wide semantic gap between the language and the object program.

However, this paper describes the successful first step toward this goal. Hardware features such as those described above can be abstracted to common high-level language features that do not make programmers conscious of the low-level hardware features. To enable this type of abstraction, a high-level language called "Phonepl" (portable high-level open network processing language) is proposed, and a method for compiling packet-handling programs in Phonepl into high-performance programs that can fully utilize hardware while distinguishing SRAM and DRAM is proposed. Here, "open" means that network processors can be programmed without NDAs. Especially, packet headers are automatically cached, the language processor is aware that the data being handled is stored in either SRAM or DRAM (or both) and manages data transmission between them, and programmers do not have to pay attention to this distinction, so the programming cost can be decreased.

Phonepl does not depend on vendor-specific NP hardware and software, and thus the programs in Phonepl can be portable among various NPs.

The rest of this paper is organized as follows. Section 2 describes related work. Section 3 describes Phonepl. Section 4 describes a method for implementing Phonepl for NPs, especially four representations of packet type and a method for handling them. Section 5 describes a prototype implementation of Phonepl for plug-ins for a network-virtualization environment called the VNode Infrastructure, and Section 6 evaluates it by using several applications. Section 7 concludes this paper.

2. Related Work

This section focuses on previous studies on NPs and languages for packet processing because, although there are many studies on memory-related optimizations concerning high-performance computing, such as Sequoia [3], they focus on array processing and the requirements for packet-stream processing are quite different from them.

2.1. Selection of SRAM/DRAM in NPs

The IXP series of NPs developed by Intel [4] does not have cache, and its SRAM and DRAM have different memory spaces. The developers at Intel reported that cache is not effective in the case of NPs, so this type of memory architecture is good for network processing. However, it is difficult to program IXP processors because programmers, who are not even aware of the difference between SRAM and DRAM, must use them with different methods.

In contrast, the architectures of NPs developed later, for example, Cavium Octeon® [5] and Tilera® Tile Processors [6], are more similar to those of general-purpose CPUs with cache. However, because a cache miss may disable wire-rate transmission of packets, there are several devices that can be applied to avoid cache miss. That is, data to be processed at wire rate must be stored in SRAM. However, because an NP cannot usually have sufficient quantity of SRAM to store all the processing packets, it only stores descriptors and headers of packets in SRAM, and the rest or whole packets must be stored in DRAM. Various different types of packet-processing hardware and software behave in a similar way. In addition, to process packets at wire rate, NPs distribute packets to many cores for parallel processing, and they sort the resulting packets by hardware in input order and queue them for output or the next processing.

2.2. Selection of SRAM/DRAM with a Packet-Processing Language

In an NP program-development environment called Shangri-La [7], which was developed by Intel and several universities, a high-level language called Baker [4] was developed for IXP. By assuming that packet bodies are stored in DRAM and descriptors are stored in SRAM, Baker enabled programmers to handle packet data without having to consider whether they are on DRAM or SRAM. The data structure on SRAM, however, must be designed by programmers, so it depends on NP architecture. In addition, programmers must describe data transmission between DRAM and SRAM, so they must explicitly describe caching operations.

Unlike Octeon or Tilera, Baker does not have a mechanism for supporting automatic distinguished use of SRAM and DRAM. It is therefore difficult to process packets at wire rate by using Baker.

2.3. Packet-Stream and Data-Stream Languages

Click [8] is software architecture for describing routers modularly. Two-level description is used in Click. The lower level, or component level, is described by C, and the higher level is described by a domain-specific language, which connects modules in several ways. Click programs can be portable, but it is difficult to get high performance from portable Click programs. NP-Click [2] is a specialized implementation of Click for IXP NPs. Modules in NP-Click are written in IXP-specific C language; there-fore, the programs are not portable.

Frenetic [9] is a language for controlling a collection of OpenFlow [10] switches. It is embedded in Python but is based on SQL. It is a declarative language and processes collections (streams) of packets instead of processing individual packets procedurally in the manner of Phonepl. Because Frenetic processes packet streams, it is very similar to CQL (Continuous Query Language) [11]. Unlike Phonepl, Frenetic can only be used to program the control plane; it cannot handle the data plane.

NetCore [12] is a rule-based language for controlling OpenFlow switches. Rules in NetCore are condition-action rules; that is, rules that match incoming packets are activated.

3. Packet-Processing Language

A high-level language called Phonepl, which solves the three problems described in the introduction, is outlined.

3.1. Basic Design of Phonepl

Phonepl is designed for wire-rate (low-level) packet-processing of any format, such as a non-IP and/or non-Ethernet format, as well as designed to be as close as a conventional programming language, *i.e.*, Java, because it should be easy to handle by Java and C++ programmers.

The reason why a new language, which is open, portable, and easy to use, is designed is explained as follows. Although it is close to conventional languages, a new language is required because it is very hard to compile a general-purpose program to high-performance object program for NPs, which very optimized hardware-usage, especially memory usage, is required for. Phonepl may thus be considered as a very restricted and extended version of Java.

Two major design goals of Phonepl are as follows. First, Phonepl must be high-level; that is, it must be designed for the programmer not to be aware of proprietary hardware and software. Second, Phonepl must be able to express high-performance packet-processing programs. Especially, processing at wire-rate, *i.e.*, 10 Gbps or more, without packet drops is required. In an NP, input packets may be partially cached, that is, the header of the packet is stored in SRAM and the rest or whole packet is stored in DRAM, but a DRAM access may disable wire-rate processing and cache miss easily cause packet-drops. However, this goal must be achieved without abandoning the first goal, *i.e.*, high-level programmability.

To achieve these design goals, data structures, especially Packet and String, which are the most important data structures in Phonepl, must be carefully designed and the method for processing them must be developed. Especially, packets are designed to be immutable byte strings in Phonepl and they are distinguished from non-packet strings.

There are five language features concerning this design. The first feature is that packets are byte strings because packets with arbitrary formats should be able to be handled in uniform methods. Packets have variable length, so they can be handled as byte strings (similar to character strings). A packet in Phonepl is not a encapsulated object. This decision makes low-level and cross-layer optimization of packets easier. The protocol-handling method written in Phonepl is thus completely different from that written in Java.

The second feature is that packets are immutable. Packets are handled as immutable (non-rewritable) objects, which are similar to character strings in Java or other languages; that is, packet contents cannot be rewritten. This immutability enables memory areas, especially DRAM areas, to be shared by packets before and after an operation.

The third feature is that types of packets, *i.e.*, Packet, and non-packet strings, *i.e.*, String, are different in Phonepl. They are incompatible for two reasons. First, although they can be logically identical, they must be implemented by using quite different methods and this distinction makes implementation more efficient and easier. Operations such as subpacket and substring described below utilize this difference. Second, programmers can easily distinguish them. Non-packet strings are used for temporary data, e.g., packet fragments, but packets are used for I/O data; that is, packets and packet fragments (non-packets) are different for programmers.

Two assumptions are made in regard to implementation of these data types. The first assumption is that whole String objects are stored in cacheable memory, *i.e.*, in SRAM, but can be stored in DRAM if needed. If they are in cache, purging the cache may have to be inhibited. The second assumption is that only the head of a packet is cached, and the tail is stored only in DRAM. However, a short packet may be wholly cached and may be stored only in SRAM.

The fourth feature is that packet and non-packet byte-substring operations are different in Phonepl because the types of the operation results are different. A new packet can be generated by removing part of another packet using a subpacket operation, and a non-packet byte string can be generated by extracting part of a packet using a substring operation. These operations can have the same name *i.e.*, a substring, but are distinguished.

The fifth feature is that packet- and byte-*concatenation* operations are specialized. A byte string can be generated by concatenating two or more byte strings by a concat operation, and a packet can be generated by concatenating one or more byte strings and a packet by a packet constructor called "new Packet". Although a packet can logically be generated by concatenating multiple packets, such concatenation seems to be practically less useful and difficult to implement, so no such operation is included (See Section 4.2.3 for more explanations).

3.2. Program Example and Packet Operations

To outline Phonepl and to explain several data structures and important packet operations, a program that performs MAC-header addition/removal, which cannot be performed by conventional non-programmable network nodes, is shown in **Figure 1**. The program in this figure defines class AddRemMAC. It has two functions that handle two bidirectional packet streams, *i.e.*, NetStream1 and NetStream2 (lines 001 - 002), which are bound to physical network interfaces outside this program. One function inputs packets from NetStream1, generates new packets with a new MAC header (*i.e.*, adds a new MAC header at the front) for each packet, and outputs them to NetStream2. The other function inputs packets from NetStream2, removes the MAC header in front, and outputs it to NetStream1. The program is much simplified because it is sufficient to show the functionality and basic implementation of the language; that is, no validation test is performed before the header is added or removed. However, it is easy to add check code to this program.

Packet flows are handled as "streams" in Phonepl. Method of stream handling is described using the constructor of class AddRemMAC here. The parameter declarations of AddRemMAC (lines 006 - 007) specify that input packets to parameter port1 pass to method process1 and input packets to parameter port2 pass to method process2. This type of parameter declaration is Phonepl specific; that is, Java grammar is modified for the sake of stream processing. The parameter values (packet streams) are assigned to instance variables out1 and out2 to make them available in the newly created object. Methods process1 and process2 receive one packet at a time. (One of these methods is executed once on only one core for each packet.) Because Phonepl handles input packets by these methods only, there is no specific method or statement for packet input.

Examples of a substring operation (which is used for accessing packet components), a packet constructor (which is used for packet composition), and a packet-stream output using "put" method can be seen in method

```
001    import NetStream1;
002    import NetStream2;

003    class AddRemMAC {
004       NetStream out1;
005       NetStream out2;

006       public AddRemMAC(NetStream port1 > process1,
007                        NetStream port2 > process2 ){
008          out1 = port1;
009          out2 = port2;
010       }

011       void process1(Packet i) {
                          //Port 1 to 2 (no VLAN -> no VLAN)
012          Packet o = new Packet(i.substring(0,14),i);
                // MAC header of original packet (i: Original packet)
013          out2.put(o);
014       }

015       void process2(Packet i) {
                          // Port 2 to 1 (no VLAN -> no VLAN)
016          Packet o = i.subpacket(14);
                // remove MAC header (no VLAN)
017          out1.put(o);
018       }

019       void main() {
020          new AddRemMAC(new NetStream1(),
021                        new NetStream2());
022       }
023    }
```

Figure 1. Simple MAC-header addition/removal program.

process1 (line 011). This method handles a packet that comes from NetStream1, generates a byte string from the first 14 bytes of input packet i (it is assumed that the size of MAC header is 14 bytes) by i.substring(0,14), generates a packet by concatenating this byte string and the original packet by new Packet(···,i), and outputs the resulting packet to NetStream2(out2).

An example of subpacket operation, which generates packets from an existing packet, can be seen in method process2 (line 015). This method handles a packet that comes from NetStream2, generates a packet by removing the first 14 bytes of input packet i by i.subpacket(14), and outputs the resulting packet to Net-Stream1 (out1).

Finally, an example of stream initialization is seen in function main() (line 019). When class AddRemMAC is initialized, this function is executed. It logically runs only once, but each processor core may execute it once unless there are side-effects. It generates an instance (a singleton) of class AddRemMAC, which runs forever and processes packets repeatedly unless it is externally terminated. Two packet streams are generated and passed as arguments of AddRemMAC. They start to operate (input and/or output packets) when instances are generated.

4. Implementation Method

To implement semantics close to conventional programming languages such as Java, a special method of handling data (object) is required for Phonepl. The key feature of Phonepl implementation is the four representations of packets and operations among them.

4.1. Four Representations of Packets

In Phonepl, multiple packet data-representations used in NPs are unified as a single data type called Packet. Four different representations shown in **Figure 2(a)** (explained below) are therefore used for Packet. These representations are required because of the following two reasons concerning high-performance packet-processing and NP hardware. First, in most packet-processing in network nodes, packet headers are added, removed, or updated, but packet tails, *i.e.*, payloads, are not touched unless very deep packet-inspection is required. So the packet headers must be stored in SRAM (or scratchpad memory) but the packet tails can be stored in DRAM as described in the introduction and in the previous section. It is usually not possible to cache whole packet. Second, NPs are designed to handle input and/or output packets by specialized hardware. The hardware is optimized for the packet-processing requirements described above, but some hardware-specific restrictions apply in addition.

An example of hardware-specific data representation that matches the abstract representation is shown here. In some NPs, there are input-specific and output-specific packet formats using a special descriptor format. Short packets may be fully stored in SRAM but packet heads may be stored in both SRAM and DRAM for longer packets. The four abstract representations are designed to generalize various concrete representations, such as shown in **Figure 2(b)**, used in NPs. Although the descriptor format is specialized, it can be abstracted as shown in **Figure 2(a)**. If vendor-specific C language is used, these representations are handled separately; however, Phonepl, handles them uniformly. Even for cases that the NP has a cache, it is probably useful to distinguish multiple representations because cache miss must be avoided.

The four representations are explained in the following.

- **Cached:** The whole packet data is stored in SRAM. It is not assumed that a copy of the data is stored in DRAM.
- **Mixed:** The head of a packet (the number of bytes depends on implementation) is stored in SRAM, and whole packet data is stored in DRAM.
- **Gathered:** A packet consists of multiple fragments. Each fragment is stored in a memory area (*i.e.*, DRAM or SRAM). A gathered packet can be represented by an array or a linked list of fragments.
- **Uncached:** The whole packet is stored in DRAM. It is not assumed that a copy of the data is stored in SRAM.

Packets inputted to NPs are usually in cached or mixed representation; that is, short packets may be represented by cached representation but mixed representation is required for long packets. All four representations are used for expressing operation results and may be used for output. However, reasoning of mixed, gathered, and uncached representations are explained more.

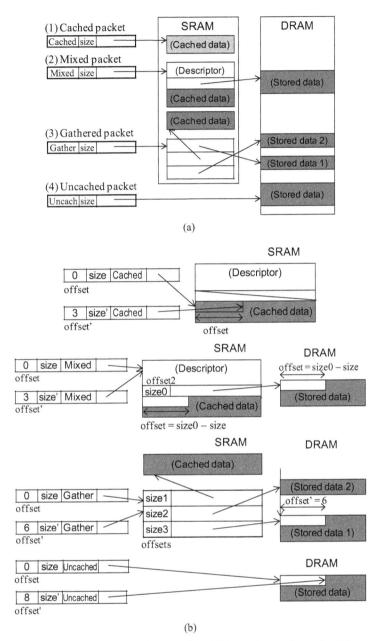

Figure 2. Four representations of packet type. (a) Abstract representations; (b) Examples of more detailed representations.

Mixed representation is required because, in packet processing, only the packet head (containing headers) is usually modified, headers are added or deleted, and the packet tail is kept unchanged. Good performance can therefore be obtained by caching only the head to SRAM and storing the tail only in DRAM. Data accessed by cores must be stored in SRAM because if data stored in DRAM is accessed, it takes excessively long time, and wire-rate processing becomes impossible.

Gathered representation is required when generating a packet from multiple pieces of data stored in DRAM or SRAM. In such a case, if all the pieces are copied to a contiguous area (of DRAM), copy from DRAM to DRAM is required and wire-rate processing becomes impossible. This representation is closely related to the immutability of packets, which enables sharing part of a string.

Uncached representation is required when a packet is generated from a tail of another packet with gathered representation by an operation such as a header deletion.

Because the four representations may have to be distinguished at run time, a tag must be supplied. The tags should be in packet-data pointers. However, because packet data are handled by hardware in NPs, the data representation and handling methods in the case of a high-level language must be very carefully designed and implemented. If the address space is sufficiently large, a part of the address can be used for a tag. This representation is close to widely used methods for dynamically-typed languages, such as Python or Lisp.

4.2. Packet Operations and Four Representations

Because there are four packet-data representations and each packet data has a tag, packet operations must be implemented for all these representations, and sometimes run-time tag check is required.

4.2.1. Run-Time Tag Check

Because there are multiple representations in Packet type, they must be distinguished dynamically (by the run-time routines in the NP) or statically (by the Phonepl compiler). In terms of efficiency, it is better for the representation to be statically distinguished. However, it is impossible to distinguish every representation of a packet statically, so run-time tag-check is, at least sometimes, necessary. Especially, if a non-optimizing compiler is used, tag check is always necessary at run time. Such a run-time check causes overhead, but it does not usually prevent wire-rate processing because the tags are in cached pointers and a tag can be added and removed with very small cost.

4.2.2. Packet I/O

Some NP hardware creates a descriptor when receiving a packet. The descriptor is in SRAM, and whole packet data may be stored in DRAM. The input packet format, thus, is close to mixed representation (or cached representation in the case of a short packet); however, a tag must be added when run-time tag-check is required. The run-time routine should thus decide which representation is to be used and insert the tag value. This means that the language processor must fill the gap (*i.e.*, convert) between data representations in the hardware and in Phonepl. If the gap is wide, significant CPU time is required to fill it, and performance may decrease. An appropriate representation design is therefore important.

An output packet format must be prepared for some NPs when sending a packet. One of the four representations should be close to the output format; however, the tag must be removed before passing the data to the packet output hardware. For example, the output format may be close to gathered representation, but the tag value "gathered" must be cleared. The hardware concatenates the fragments pointed to by the gathered representation and outputs the result.

4.2.3. Subpacket

Each representation requires different implementations of an operation to achieve a subpacket operation. In all the cases described below, the operations are executed using data stored in SRAM, and DRAM is not accessed.

If the packet has a cached representation, a subpacket of the packet is in a cached format. The original packet can be stored in the allocated SRAM area. The resulting subpacket may share the original packet data or may be a copy of the original data. In this case, because both the original and copied data are stored in SRAM, this copy operation probably does not prevent wire-rate processing.

If the packet has a mixed representation, a subpacket of the packet may be in a mixed or uncached format. That is, there are two cases. Firstly, if the resulting packet contains both head data stored in SRAM and tail data stored in DRAM, the result is mixed format. Secondly, if the resulting packet only contains tail data, the result is uncached format. In general, the resulting representation is not known at compile time because the range specified in subpacket operation might not be known at compile time. In both cases, a new descriptor is generated in SRAM by using the original descriptor, but no packet data stored in DRAM is accessed.

If the packet has a gathered representation, a substring of the packet is usually in a gathered format. The original and resulting packets may share the array of fragments (*i.e.*, only a packet-type pointer is generated) or the resulting pointer may point to a new array copied from the original array. An array copy probably does not prevent wire-rate processing because both arrays are stored in SRAM.

If the packet has an uncached representation, a substring of the packet is in an uncached format. Both the original and resulting packet data are stored in DRAM and shared. The address and the length of the resulting packet are stored in a packet-type pointer. No packet data stored in DRAM are accessed.

4.2.4. Concatenation

When a packet is generated by concatenating one or more byte strings (such as new headers and a packet content), a constructor, "new Packet()", is used. In the current implementation method, this constructor generates a gathered-format packet. That means, the parameter values of the constructors are the elements of the array in the gathered format. However, a more optimized method, which uses other representations, may be developed.

The last element of the constructor may be a packet of any representation. If this element has a mixed format, the DRAM part (which represents the whole packet) becomes an element of the array. If this element has a gathered format, each input array element becomes an element of the array of the output gathered format.

4.2.5. Generating Packet without Using Input Packet

A packet can be created without using a pre-existing packet by using a packet constructor. The generated packet is in cached or gathered format. If the constructor has only one argument that contains a byte string, the resulting packet is in cached format, and if it has two or more arguments, the resulting packet is in gathered format.

4.3. Several Miscellaneous Issues

Two issues related to the proposed packet-handling method are explained in the following. The first issue is memory deallocation. Sharing part of packets and strings makes memory deallocation difficult. Garbage collection or reference counting can solve this problem completely, but the overhead is large. In the current implementation, strings that are (potentially) assigned to global (instance) variables are not deallocated. However, the current deallocation policy may cause memory leak. A more precise method should be devised in future work.

The second issue is adaptation to hardware-based memory allocation. Some NPs allocate and deallocate packet memory automatically to avoid software-memory-management overhead. When a packet arrives, the SRAM and DRAM required for the packet is allocated. However, it is difficult for NP hardware to decide when the packet memory can be deallocated. A Phonepl compiler must therefore generate code for deallocate it.

5. Prototyping

The above-described implementation method has been applied to a programming environment called +Net, which contains a Phonepl processor called +Net Phonepl. +Net Phonepl is used for programming physical nodes with a network-virtualization function and NPs.

5.1. Platform

The prototype compiles a Phonepl program and runs it on a "virtualization node" (VNode) [1] [13]. A virtualization platform called VNode Infrastructure, which supports multiple slices (i.e., virtual networks) using a single network infrastructure, and a high-performance fully functional virtualization testbed were developed. The components of a VNode contain NPs. The prototype is a replacement of one or more NPs in this environment. The program has packet I/O streams as described in Section 3.

A source program is compiled according to the following procedure. First, an intermediate language program (ILP) is generated by using a Phonepl syntax/token analyzer. The syntax analyzer was generated using "Yet Another Perl Parser" (YAPP) compiler, which has similar functions as those of YACC (Yet Another Compiler Compiler) or Bison parser-generators but is written in and generates Perl code. The ILP is translated by using a Phonepl translator into a specialized C program. A GNU C compiler for Octeon compiles this C program and generates object code for an Octeon board called WANic-56512 developed by General Electric Company. A run-time library is linked to the object program. The main components of this library are an initializer, packet processors, and a packet-output routine.

5.2. Compiled Code of +Net-Phonepl Compiler

To outline the object-code structure and the compilation (or program transformation), an example of compiled code is explained here. The C program generated by the Phonepl compiler from the MAC-header addition/removal program (in **Figure 1**) is shown in **Figure 3**.

```
// Translated Code for Octeon 58XX (WANic 56582) by Phonpl Translator

#include <stdio.h>
#include <string.h>
#include "runtime.h"
#include "cvmx-helper.h"

// Packet handler vector:
void (*__packetHandler[17])(__Packetp p);

// Stream data type for packets:
typedef int NetStream;

// Method NetStream.put(uint64_t port, Packet outp)
extern int NetStream_put(uint64_t port, __Packetp outp);
```

// Omitted

```
// Instance variables of the singleton instance (Singleton assumed!)
typedef struct {                    (1) Derived from instance variable
  NetStream out1;
  NetStream out2;                        (out1, out2) declaration
} AddRemMAC;
```

```
AddRemMAC __self;
```

```
AddRemMAC* AddRemMAC_new(NetStream port1, NetStream port2);
```

```
// Method AddRemMAC.process1  (2) Derived from void process1(...)
void AddRemMAC_process1(__Packetp i) {
  __Packetp o = __Packet_concat2(__Packet_substring(i, 0, 14), i);
NetStream_put(__self.out2, o);
}
```

```
// Method AddRemMAC.process2  (3) Derived from void process2(...)
void AddRemMAC_process2(__Packetp i) {
  __Packetp o = __Packet_subpacket(i, 14);
NetStream_put(__self.out1, o);
}
```

```
// Constructor AddRemMAC
AddRemMAC* AddRemMAC_new(NetStream port1, NetStream port2) {
int i;
for (__i = 0; __i < 17; __i++) {          Generating a method table
  __packetHandler[__i] = 0;
}
__packetHandler[port2] = &AddRemMAC_process2;
__packetHandler[port1] = &AddRemMAC_process1;
__self.out1 = port1;
__self.out2 = port2;               (4) Derived from the constructor
return &__self;
}                                       (Public AddRemMAC(...))
```

```
// Main loop (Scheduler)
int __mainLoop(int no_ipd_wptr) {
  cvmx_wqe_t *wqe = NULL;
                                    (5) Derived from void main()
  // Omitted                            (scheduler)

  wait_for_link_up();

  // Omitted
```

```
AddRemMAC_new(0, 16);                    ─AddRemMAC object creation
```

```
for (;;) {                        Repeating the following process
  wqe = get_input_packet();
  if (wqe != NULL) {              ─for each packet (in wqe)
```

// Omitted

```
  __Packetp __wqep;                  Phonepl packet-pointer creation
  __wqep.u64 = 0;
  __wqep.s.pool = CVMX_FPA_WQE_POOL;
  __wqep.s.size = wqe->len;
  if (wqe->word2.s.bufs == 0) {
                /* if no buffered data (no data in DRAM) */
    __wqep.s.addr = cvmx_ptr_to_phys(wqe->packet_data);
    // *** IPv4/v6 cases? ***
    Set_packet_representation(__wqep, CSP_CACHED);
  } else {          /* if data both in DRAM and in cache */
    __wqep.s.addr = cvmx_ptr_to_phys(wqe);
    Set_packet_representation(__wqep, CSP_MIXED);
  }                                              Tag insertion
```

```
  if (__packetHandler[wqe->ipprt]) {
    (*__packetHandler[wqe->ipprt])(__wqep);
  }
                                  ─Processing a packet by calling
  }                                AddRemMAC_process1 or
  return 0;                        AddRemMAC_process2
}
```

Figure 3. Compiled code of MAC-header insertion/deletion program.

This program is explained instead of describing detailed compilation process because the process is too much complicated and the program structure can probably be used for other types of NPs. A compilation technique specialized for a singleton (*i.e.*, single-instance class) is applied to this program. Cores in an Octeon processor execute this program in parallel; that is, each core processes a packet. The program consists of five parts: part 1 derived from instance-variable declaration, parts 2 and 3 derived from methods process1 and process2, part 4 derived from the constructor, and part 5 derived from the main program.

In part 1, AddRemMAC type, which corresponds to instance of class AddRemMAC in Phonepl, is declared. Because a compiled object of class AddRemMAC has two objects of NetStream type, the corresponding structure components are declared. In parts 2 and 3, *i.e.*, method definitions, the element names in the source program are replaced by the element names in the run-time library. The run-time routines may be expanded in-line; however, they are not expanded in this example program. In part 4, *i.e.*, the constructor of class AddRemMAC, methods process1 and process2 are initialized. Assignment statements that correspond to the assignment statements in the source program are included in this part. In part 5, *i.e.*, the main program, the above constructor is called, and every time it receives a packet, one of the above two methods are called. Because NetStream type is an abstraction of a packet stream, the stream elements are handled one by one, and the scheduler for this process occupies the main part of part 5. When the function get_input_packet() is called, a packet is received, and the data representation of this packet is converted to that of +Net Phonepl by adding a tag, *i.e.*, cached (CSP_CACHED) or mixed (CSP_MIXED).

6. Evaluation

Both the programmability, especially ease of language use, and the performance of the implementation should be evaluated; however, because Phonepl is being improved, performance is focused in this evaluation. Two Phonepl programs for network-layer packet handling were written. Prototypes with these object programs were used for extending VNode, and the traffic was measured.

6.1. MAC-Header Addition/Deletion Program

The first program performs MAC-header addition/removal. It is a modified version of the program shown in **Figure 1**, and similar programs are used for extending virtualization-node (VNode) functions by using the node plug-in architecture [14]-[16]. Instead of duplicating the MAC header, the Phonepl program inserts a constant MAC header that contains fixed source and destination MAC addresses and a TEB type value (*i.e.*, transparent Ethernet bridge, x6558).

As shown in **Figure 4**, the above program was used in an extended VNode, which is a gateway between slices and external networks and is called NACE or NC [17]. This network consists of the VNode and two personal computers, PC1 and PC2. PC1 simulates a terminal or a virtual node in a slice. PC2 is in an external physical network. The VNode connects the slice and the external network, and it must convert the packet format, *i.e.*, convert from the internal to external protocols, and vice versa, but the base component of the VNode does not have this conversion function. The VNode is experimentally extended by the node plug-in architecture with the

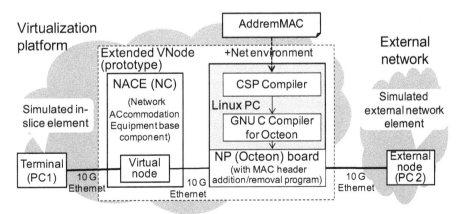

Figure 4. Extended VNode environment for experiments.

+Net environment, which consists of a PC with a Phonepl compiler, run-time routines, a GNU C compiler for Octeon, and WANic-56512 with twelve-core 750-MHz Octeon. By using conversion programs written in Phonepl, the VNode can adapt to various types of external networks.

Maximum performance of the test program was measured by using a network-measurement-tool suite called IXIA. Both operations, *i.e.*, MAC-header addition and deletion, were measured, and compared with a pass-through program, which is also written in Phonepl. The measurement results are shown in **Figure 5**. In this experiment, the input packet representation is mixed or cached, and the output packet representation is mixed or cached for header deletion and it is gathered for header addition. Uncached format is not used here because not whole cached data is removed by the header deletion. The maximum throughput (input rate) that can be passed with almost no packet drop is over 7.5 Gbps when the packet size is 256 bytes or larger. This throughput is close to the wire rate. The throughputs of two programs are mostly the same, indicating that the major overhead lies in the hardware or the initialization/finalization code, namely, not in the compiled code or the packet/string run-time routines.

Table 1 compares the performance of the Phonepl program on the Octeon and a sequential C program on eight-core 3-GHz Intel Xeon processors. Although the performance of the former is much higher, it is mainly caused by the number of used cores. If all the cores are used, the throughput of Xeon may be better; however, it is very hard to use multiple cores and to preserve the order of packets in Xeon. As shown in **Table 1**, the Phonepl program is much shorter even when compared with the C program.

Moreover, **Table 1** suggests an important difference between the two implementations; that is, the packet loss ratio is slowly increasing in the Xeon implementation because cache miss is slowly increasing, but packets are almost never lost if the input ratio is 9.2 Gbps or less in the Phonepl implementation because the memory usage is completely controlled.

6.2. Timestamp Handler for Network Virtualization Platform

The second program, which is described in detail in another paper [18], is a program for measuring communication delay between two points in the virtualization network. In this evaluation, NPs and the program was used only in VNodes, and a slow-path program was used in the gateways.

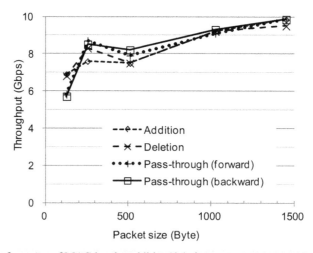

Figure 5. Performance of MAC-header addition/deletion.

Table 1. Results of MAC header addition/deletion.

Implementation	Throughput (Gbps)*		Program lines
	Header addition	Header deletion	
Phonepl program	9.2†	9.2†	26‡
C program (Xeon, single core)**	2.3† (4.0††)	1.7† (4.0††)	161‡

*Packet size: 1024 B; **Promiscuous mode is used; †No packet loss (ratio $< 10^{-6}$); ††Packet loss ratio $= 10^{-3}$; ‡Comment-only lines are not counted.

A VNode platform can support delay measurement function without adding programs and data (*i.e.*, packet format) for measurement to programs in virtual nodes. This function is useful when slice developers want to measure delay of a high-bandwidth application with certain intelligent functions in relaying nodes. A special type of virtual links between nodes, which is called measurable VLAN virtual link (MVL) type and developed by using the VNode plug-in architecture, is used to implement this function. MVLs are implemented by using timestamp insertion/deletion programs in the nodes. A VNode removes the platform header, which includes a GRE/IP or VLAN header and the timestamp, from an incoming packet and adds one to an outgoing packet, so programs that handles packets on a slice never see the platform header.

The virtualization-network structure used for this experiment is drawn in **Figure 6**. Two terminals communicate using a slice. The physical network contains two VNodes. Each VNode contains a virtual node, which are connected by an MVL. In the platform, each packet has a platform header with a timestamp.

The communication and measurement methods used for this experiment is as follows. The timestamp is inserted at the entrance gateway. Each VNode generates a packet for the virtual node by removing the platform header from an incoming packet and restores the timestamp to outgoing packets that comes from the virtual node and are identified with a stored incoming packet. The timestamp is tested and deleted at the exit gateway, which calculates the delay between the entrance and exit gateways. In the network described in **Figure 6**, the two VNodes and one PC is used for the two gateways (and terminals) to avoid the difficult synchronization problem. Terminal PCs communicate each other by using Ethernet packets, which are switched by the MAC addresses in the virtual nodes. A WANic-56512 that contains the program handles both incoming and outgoing packets. An Ethernet switch program, which is a slow-path program, works on a virtual node in a VNode.

The NP also swaps the external and internal MAC addresses in the platform header [1]. To swap addresses, the program contains a conversion table for these MAC addresses, which is implemented using a string array, and accepts virtual-link-creation and deletion requests. A creation request adds an entry to the conversion table.

The results show the gateway-to-gateway delay is 178 μS (σ = 24 μS). **Table 2** compares the performance of the 750 MHz Octeon and the 3-GHz Xeon. The performance is very close to wire rate. The C program is relatively short because this program does not contain conversion-table configuration code but the Phonepl program contains it. However, the former is still much longer.

7. Concluding Remarks

An open, portal, and high-level language, called Phonepl, is proposed. By using Phonepl, a programmer can develop a program that uses SRAM and DRAM appropriately without having to be aware of a distinction between SRAM and DRAM. To handle packets appropriately in this environment, four packet data-representations and packet-operation methods are proposed. A prototype using Octeon NP was developed and evaluated. The

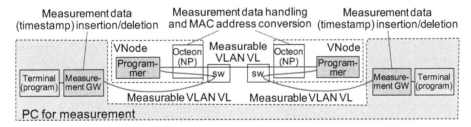

Figure 6. Virtualization-network structure for time-stamp handling.

Table 2. Results of timestamp handling and conversion.

Implementation	Throughput (Gbps)*		Program lines
	Header addition	Header deletion	
Phonepl program	10.0[†]	9.5[†]	99[‡]
C program (Xeon, single core)**	2.3[†] (4.0[††])	2.2[†] (4.0[††])	190[‡]

*Packet size: 1024 B; **Promiscuous mode is used; [†]No packet loss (ratio < 10^{-6}); [††]Packet loss ratio = 10^{-3}; [‡]Comment-only lines are not counted.

throughput of the prototype system is close to the wire rate, *i.e.*, 10 Gbps, when the packet size is 256 bytes or larger, in several packet-conversion applications. Although this is a preliminary result, it proves the proposed method is promising in achieving our objectives, *i.e.*, popularity among developers, reduced cost in programmability, and portability.

Future work includes evaluation of Phonepl language and processor by human programmers and improvement of the language design and implementation according to the evaluation result. Although Phonepl and the language processor inevitably have limitations, they should be acceptable and, if possible, natural to programmers. Future work also includes implementation of Phonepl for other types of NPs to prove the portability. Moreover, the memory allocation and deallocation mechanism must be improved to reduce memory leak caused by global variable assignments and the performance of the Phonepl language processor should be improved.

Acknowledgements

The author thanks Professor Aki Nakao from the University of Tokyo, Yasushi Kasugai, Kei Shiraishi, Takanori Ariyoshi, Takeshi Ishikura, and Toshiaki Tarui from Hitachi, Ltd., and other members of the project for discussions and evaluations. Part of the research results described in this paper is an outcome of the Advanced Network Virtualization Platform Project A funded by National Institute of Information and Communications Technology (NICT).

References

[1]   Kanada, Y., Shiraishi, K. and Nakao, A. (2012) Network-Virtualization Nodes That Support Mutually Independent Development and Evolution of Components. *IEEE International Conference on Communication Systems* (*ICCS* 2012). http://dx.doi.org/10.1109/iccs.2012.6406171

[2]   Shah, N., Plishker, W., Ravindran, K. and Keutzer, K. (2004) NP-Click: A Productive Software Development Approach for Network Processors. *IEEE Micro*, **24**, 45-54. http://dx.doi.org/10.1109/mm.2004.53

[3]   Fatahalian, K., Knight, T.J., Houston, M., Erez, M., Horn, D.R., Leem, L., Park, J.Y., Ren, M., Aiken, A., Dally, W. J. and Hanrahan, P. (2006) Sequoia: Programming the Memory Hierarchy. 2006 *ACM/IEEE Conference on Supercomputing*. http://dx.doi.org/10.1109/sc.2006.55

[4]   Goglin, S.D., Hooper, D., Kumar, A. and Yavatkar, R. (2003) Advanced Software Framework, Tools, and Languages for the IXP Family. *Intel Technology Journal*, **7**, 64-76.

[5]   Cavium Networks (2010) OCTEON Programmer's Guide. The Fundamentals. http://university.caviumnetworks.com/downloads/Mini_version_of_Prog_Guide_EDU_July_2010.pdf

[6]   Bell, S., Edwards, B., Amann, J., Conlin, R., Joyce, K., Leung, V., MacKay, J., Reif, M., Bao, L.W., Brown, J., Mattina, M., Miao, C.-C., Ramey, C., Wentzlaff, D., Anderson, W., Berger, E., Fairbanks, N., Khan, D., Montenegro, F., Stickney, J. and Zook, J. (2008) TILE64-Processor: A 64-Core SoC with Mesh Interconnect. *IEEE International Solid-State Circuits Conference* (*ISSCC* 2008), San Francisco, 3-7 February 2008, 88-598. http://dx.doi.org/10.1109/isscc.2008.4523070

[7]   Chen, M.K., Li, X.F., Lian, R., Lin, J.H., Liu, L.X., Liu, T. and Ju, R. (2005) Shangri-La: Achieving High Performance from Compiled Network Applications While Enabling Ease of Programming. 2005 *ACM SIGPLAN Conference on Programming Language Design and Implementation* (*PLDI'*05), 224-236. http://dx.doi.org/10.1145/1064978.1065038

[8]   Kohler, E., Morris, R., Chen, B.J., Jannotti, J. and Frans Kaashoek, M. (2000) The Click Modular Router. *ACM Transactions on Computer Systems*, **18**, 263-297. http://dx.doi.org/10.1145/354871.354874

[9]   Foster, N., Harrison, R., Freedman, M.J., Monsanto, C., Rexford, J., Story, A. and Walker, D. (2011) Frenetic: A Network Programming Language. 16*th ACM SIGPLAN International Conference on Functional Programming* (*ICFP'*11). http://dx.doi.org/10.1145/2034773.2034812

[10] McKeown, N., Anderson, T., Balakrishnan, H., Parulkar, G., Peterson, L., Rexford, J., Shenker, S. and Turner, J. (2008) Open Flow: Enabling Innovation in Campus Networks. *ACM SIGCOMM Computer Communication Review*, **38**, 69-74. http://dx.doi.org/10.1145/1355734.1355746

[11] Arasu, A., Babu, S. and Widom, J. (2006) The CQL Continuous Query Language: Semantic Foundations and Query Execution. *The VLDB Journal*, **15**, 121-142. http://dx.doi.org/10.1007/s00778-004-0147-z

[12] Monsanto, C., Foster, N., Harrison, R. and Walker, D. (2012) A Compiler and Run-Time System for Network Programming Languages. *Proceedings of the* 39*th ACM SIGPLAN-SIGACT Symposium on Principles of Programming Languages*, Philadelphia, 25-27 January 2012, 217-230. http://dx.doi.org/10.1145/2103656.2103685

[13] Nakao, A. (2012) VNode: A Deeply Programmable Network Testbed through Network Virtualization. *Proceedings of*

the 3rd IEICE Technical Committee on Network Virtualization, Tokyo, 2 March 2012. http://www.ieice.org/~nv/05-nv20120302-nakao.pdf

[14] Kanada, Y. (2013) A Node Plug-In Architecture for Evolving Network Virtualization Nodes. 2013 *Software Defined Networks for Future Networks and Services* (*SDN4FNS*). http://dx.doi.org/10.1109/sdn4fns.2013.6702531

[15] Kanada, Y. (2014) A Method for Evolving Networks by Introducing New Virtual Node/Link Types Using Node Plug-Ins. *Proceedings of the IEEE Network Operations and Management Symposium* (*NOMS*), Krakow, 5-9 May 2014, 1-8. http://dx.doi.org/10.1109/noms.2014.6838417

[16] Kanada, Y. (2014) Controlling Network Processors by Using Packet-Processing Cores. *Proceedings of the 28th International Conference on Advanced Information Networking and Applications Workshops* (*WAINA*), Victoria, 13-16 May 2014, 690-695. http://dx.doi.org/10.1109/waina.2014.112

[17] Kanada, Y., Shiraishi, K. and Nakao, A. (2012) High-Performance Network Accommodation into Slices and In-Slice Switching Using a Type of Virtualization Node. *2nd International Conference on Advanced Communications and Computation* (*Infocomp* 2012), IARIA.

[18] Kanada, Y. (2014) Extending Network-Virtualization Platforms Using a Specialized Packet-Header and Node Plug-Ins. *Proceedings of the 22nd International Conference on Telecommunications and Computer Networks*, Split, 17-19 September 2014.

4

Improving Queuing System Throughput Using Distributed Mean Value Analysis to Control Network Congestion

Faisal Shahzad[1], Muhammad Faheem Mushtaq[1], Saleem Ullah[1*], M. Abubakar Siddique[2], Shahzada Khurram[1], Najia Saher[1]

[1]Department of Computer Science & IT, The Islamia University of Bahawalpur, Bahawalpur, Pakistan
[2]College of Computer Science, Chongqing University, Chongqing, China
Email: faisalsd@gmail.com, faheem.mushtaq88@gmail.com, *saleemullah@iub.edu.pk, abubakar.ahmadani@gmail.com, khurram@iub.edu.pk, najia@iub.edu.pk

Abstract

In this paper, we have used the distributed mean value analysis (DMVA) technique with the help of random observe property (ROP) and palm probabilities to improve the network queuing system throughput. In such networks, where finding the complete communication path from source to destination, especially when these nodes are not in the same region while sending data between two nodes. So, an algorithm is developed for single and multi-server centers which give more interesting and successful results. The network is designed by a closed queuing network model and we will use mean value analysis to determine the network throughput (β) for its different values. For certain chosen values of parameters involved in this model, we found that the maximum network throughput for $\beta \geq 0.7$ remains consistent in a single server case, while in multi-server case for $\beta \geq 0.5$ throughput surpass the Marko chain queuing system.

Keywords

Network Congestion, Throughput, Queuing System, Distributed Mean Value Analysis

1. Introduction

Networks where a communication path between nodes doesn't exist refer to delay tolerant networks [1] [2]; they

---

*Corresponding author.

communicate either by already defined routes or through other nodes. Problem arises when network is distributed and portioned in to several areas due to the high mobility or when the network extends over long distances and low density nodes. The traditional approach [3] for the queuing system was to design a system of balance equations for the joint property of distributed vector value state. The traditional approaches to the system of Markovian [4] queuing systems were to formulate a system of algebraic equations for the joint probability distributed system vector valued state, which was the key step introduced by Jackson [5], that for a certain types of networks the solutions of the balance equation is in the form of simple product terms. All remained to be normalized numerically to form the proper probability distribution In case of networks with congested routing chains, this normalization turned out to be limited and degrades the system efficiency as well.

Therefore, in practical these distribution contains an extra detail, such as mean queue size, mean waiting time and throughput is needed. The framework of conventional algorithm shows that these properties can be obtained by normalizing constants. The proposed algorithm given in this paper correlates directly with the required statistics. Its complexity is asymptotically is almost equal to the already defined algorithms, but the implementation of program is very simple.

Choosing the right queuing discipline and the adequate queue length (how long a packet resides in a queue) may be a difficult task, especially if your network is a unique one with different traffic patterns. Monitoring of the network determines which queuing discipline is adequate for the network. It is also important to select a queuing length that is suitable for your environment. Configuring a queue length that is too shallow could easily transmit traffic into the network too fast for the network to accept, which could result in discarded packets. If the queue length is too long, you could introduce an unacceptable amount of latency and Round-Trip Time (RTT) jitter. Program sessions would not work and end-to-end transport protocols (TCP) would time out or not work.

Because queue management is one of the fundamental techniques in differentiating traffic and supporting QoS functionality, choosing the correct implementation can contribute to your network operating optimally.

2. Mean Value Evaluation

Included in the Quality of Service Internetwork architecture is a discipline sometimes called queue management. Queuing is a technique used in internetwork devices such as routers or switches during periods of congestion. Packets are held in the queues for subsequent processing. After being processed by the router, the packets are then sent to their destination based on priority.

In the queuing network, the traditional approach to find the solution is using characteristics of a continuous time markov chain to formulate a system of balance equation for the joint probability distribution of the system state. The solution of the balance equations, for certain classes of networks such as Jackson networks and Gordon-Newell networks is in the form of a product of simple terms, see [6]. In general, the joint probability is not so simple and sometime it is inefficient to pursue. If we only interested in the average performance measure such as average waiting time, average response time or network throughput, we do not need the steady-state probabilities of the queue length distributions. In this stage, we will use an approach so called mean value analysis. The algorithm for this approach works directly to the desired statistics and has been developed and applied to the analysis of queuing networks, for example see [7]. The mean value analysis is based on the probability distribution of a job at the moment it switches from one queue at a server to another and the steady state probability distribution of jobs with one job less. This relation is known as arrival theorem for closed queuing networks. In a closed queuing network, the bottleneck is the queue with the highest service demand per passage.

Using the arrival theorem, if a job move from queue i to queue j in a closed queuing networks with K jobs in it, will find on average $E\left[N_i \left(K-1 \right) \right]$ jobs. With this result and assuming that a job is served in a first come first served basis, a relation between the average performance measure in this network with K jobs and $K-1$ job can be performed recursively.

3. Proposed Mechanism

Mean value analysis depends on the mean queue size and mean waiting time. This equation applied to each routing chain and separately to each service center will furnish the set of equations which will easily solved numerically. The proposed algorithm is simple and avoid overflow, underflow actions which may arise with traditional algorithms. All mean values in the algorithm are calculated in a parallel manner. Thus memory requirement is higher than the previous ones, but new mechanism is relatively faster in multi-server scenarios.

We have considered the closed multi-chain queuing system which has the product form solution. Suppose C is a routing chain and S is a service center. Each chain contains a fixed number of customers who processed through subset of services using Markov chain technique, while service providers adopt one of the following mechanisms.

1) FIFO: customers are serviced in order of arrival, and multi-servers can be used.

2) Priority Queuing: customers are serviced according to the traffic categorization.

3) WFQ (Weighted Fair Queuing): gives low-volume traffic flows preferential treatment and allows higher-volume traffic flows to obtain equity in the remaining amount of queuing capacity. WFQ tries to sort and interleave traffic by flow and then queues the traffic according to the volume of traffic in the flow.

4) PS: Customers are served in parallel by a single server.

5) LCFSPR: customers are served in reverse order of arrival by a single server, (Last come first served preemptive resume).

Now we assume that all the servers have constant service rate using multiple FCFS service centers starting with the following consequences, which relates mean waiting time $W(K)$ to the mean queue size of the system $W(K - Er)$ with one customer less in the chain r making the following equation,

$$T_{r,l} = \sigma_{r,l} \{1 + n_l (K - E_r)\} \tag{1}$$

$T_{r,l}$ is the equilibrium mean waiting time of chain r at service center l $\sigma_{r,l}$ is the mean value of the service demand, n_l represent equilibrium mean queue size, K is size of chain r, and E_r is the R-dimensional unit vector. From the definition it is clear that

$$T_{r,l} = \sigma_{r,l} \tag{2}$$

K_r is the average number of customers in chain between successive visits, then the mean number of visits (θ) and its waiting time per visit with service center visited by chain $S(r)$ is described as

$$\alpha_r = k_r \left[\frac{1}{\sum_{l \in S(r)} (\theta_{r,l} T_{r,l})} \right] \tag{3}$$

where α_r is the throughput of chain r, considering service center $l(r)$ i.e. $\alpha_r = \alpha_{r,l(r)}$ then for multiple service stations each one yields the following relation,

$$N_{r,l} = \alpha_{r,l} T_{r,l} = \alpha_r \theta_{r,l} T_{r,l} \tag{4}$$

The above equations applicable for recursive analysis of mean queuing size, meat waiting time and system throughput. The initial point can be set as, $N_{r,l}(0) = 0$ For all $r = 1, \cdots, R$ and $l = 1, \cdots, L$.

To make the substitution in an algorithmic form, we have $T'_{r,l} = \theta_{r,l} T_{r,l}$ and $N'_l = 1 + N_l$.

3.1. Mathematical Model

Our model is defined on a one-dimensional closed system consisting of M cells i.e. **Figure 1**. A closed queuing network model is justified for steady state conditions. In steady state, for a single-entry and single-exit lane, the traffic flow into the system will be equal to the traffic flow out of the system. We approximate this as a closed system where the number of vehicles remains the same. Each cell can either be empty or occupied by one vehicle. To start with, we assume vehicles of identical size. Since the system is closed, the number of vehicles remains constant, say equal to N. Thus the system density can be defined as $N/M = \rho$. Hereon, this model will be referred to as the Path Cell Network model. A vehicle moves from the first to the second and so on to the Mth cell and then back to the first cell. It is apparent that the system has attributes of a queuing system with FIFO discipline. In the past, the general modeling of traffic using Queuing Theory has been macroscopic, but here instead of treating the whole closed link as a single queue, we consider it as a network of queues. The Path Cell Network model at first appears to be a cumbersome one as each cell has limited space and, therefore, each queue in the network limited buffer space. But, our task can be made much easier by our definition of the servers. The exact working of the model is as follows:

Being a single lane model each vehicle moves to the next cell if empty or waits, and then moves when the vehicle ahead vacates the cell. Thus there can only be two configurations for a particular vehicle: either the cell

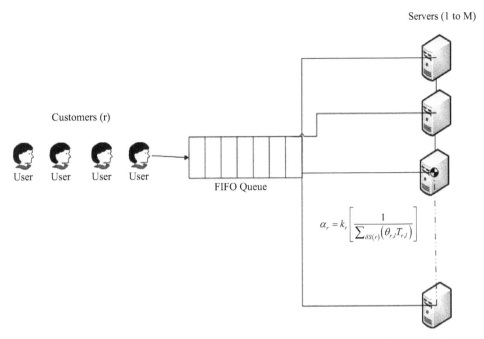

Figure 1. Queuing model.

ahead is empty or occupied. We say a vehicle is in service when the cell ahead is empty and "waiting" if it is occupied.

$$\begin{array}{l} \text{if } \left(\text{cell} = \text{occupied} \right) \\ \qquad \text{wait} \\ \text{else} \\ \qquad \text{In\_service} \end{array} \Bigg]$$

In effect, each empty cell acts as a server, and at any point in time there are always $M - N$ servers in the system that keeps changing their positions. These dynamic cells act as servers to $M - N$ queues in the system that together form a closed network. The number of waiting units in each queue can be counted as the total number of vehicles between the empty cell and the one empty behind it. Thus if there are two consecutive empty cells, both act as servers with one of them having zero queue size. Service in each queue is assumed to be exponential, and for the basic Path Cell Network model, assuming identical vehicles, the service rate of each vehicle is also taken to be the same. As the service is exponential, from the Poisson-in-Poisson-out property inter-arrival times at each queue are also exponential.

The model we claim can be mapped onto a cyclic Jackson network with $M - N$ cells and N customers. Thus, effectively, a path segment with very limited buffer space at each queue is mapped onto a well-known cyclic queuing network with buffer space of size N.

The results of a cyclic Jackson network are well known. A state is indicated by

$$k_1 k_2 k_3 \cdots k_n, \text{ where } \sum k_i = N$$

and k_i indicates the number of units at each stage of the closed queuing network. The probability of being in a state $k_1 k_2 k_3 \cdots k_n$ is written as $p\left(k_1 k_2 k_3 \cdots k_n \right)$ Transitions between states occur when a unit enters or leaves a stage. Service rate μ_i at each stage is allowed to be dependent upon the number of units i in the stage. Then

$$p\left(k_1 k_2 k_3 \cdots k_n \right) = p\left(N, 0, 0, 0 \right) \frac{\mu_1^{N - k_1}}{\mu_2^{k_2} \mu_2^{k_3} \cdots \mu_k^{k_n}} \tag{5}$$

while, $\sum \left(k_1 k_2 k_3 \cdots k_n \right) = 1$

In this case we have consider the service rate and probability at each node and stage is same and equal to the inverse of the number of ways of selecting N out of $N + k - 1$ places as customers and remaining $k - 1$ places being the partitions, it is calculated as

$$p = \frac{N!(k-1)}{k+N-1},$$

here throughput α_j of the network can be calculated as

$$\alpha_j = \mu p(k_1 > 0) = \mu(1 - p(k_1 = 0)) \tag{6}$$

where $p(nl = 0)/p$ is simply the number of ways of selecting N out of the last $N + k - 2$ places as customers, the first place being a partition

$$\alpha_j = \mu \frac{N}{N + k - 1} \tag{7}$$

For the corresponding network number of queues $k = M - N$, throughput β_r of the network is obtained by scaling α_j using the following relation

$$M_{\beta_r} = (M - N)/\alpha_j \tag{8}$$

After scaling, throughput becomes

$$\beta_r = \mu \frac{N(M-N)}{(M-1)M} \tag{9}$$

For $N, M \gg 1$, $\beta_r = \mu p(1-p)$

The algorithm starts with an empty network (zero customers), then increases the number of customers by 1 until it reaches the desired number of customers of chain r.

The average waiting time in this closed queuing network and the average response time per visit are given by the following formulas:

$$E[w_i(k)] = E[N_i(k-1)]E(s_i) \tag{10}$$

where, s_i is the service time, and w_i is waiting time. To obtain the average response time per passage, the above equation can be obtained by each time visit ratio as

$$E[P_i(k)] = E[N_i(k-1)+1]E(s_i)v_i \tag{11}$$

where, p_i is response time and v_i is each time visit. The expected number of jobs in queue I is given by little's formula [6] as.

$$E[N_i(k)] = \beta(k)E[p_i(k)] \tag{12}$$

3.2. Case Studies

Now we have to implement our model in single and multi-server scenarios to calculate throughput and mean waiting time.

a) SINGLE SERVER CASE

Initialize $N'_l[0] = 0$ for all $l = 1, 2, 3, \cdots, L$

$$\text{for } (i_1 = 0, \cdots, K_1), \text{ for } (i_2 = 0, \cdots, K_2), \text{ and } \text{ for } (i_R = 0, \cdots, K_R)$$

If at the service centers customers are delayed independently (D) of other customers, then we have the following steps,

$$\left. \begin{array}{l} \text{if}\left(Service\_Center == D\right) \\ T'_{r,l} = \Upsilon_{r,l}\delta\left(i_r\right) \\ \text{else} \\ T'_{r,l} = \Upsilon_{r,l}N'_l\left(i-e_r\right) \end{array} \right\} \quad \forall r = 1,2,\cdots,R \tag{13}$$

Then we have a little's equation for chains and service centers as,

$$\alpha_r = \frac{l_r}{\displaystyle\sum_{l\in S(r)} T'_{r,l}}, \quad \forall r = 1,\cdots,R \tag{14}$$

$$T'_l(I) = 1 + \sum_{r\in R(l)} \alpha_r T'_{r,l}, \quad \forall l = 1,2,\cdots,L \tag{15}$$

The operations count for this algorithm is bounded by $2RL - R$ additions and $2RL + R$ multiplication/divisions. This is the same as the convolution algorithm in its most efficient. However, Algorithm 1 completely avoids a genuine problem of the convolution algorithm, namely, that the floating point range of many computers may be easily exceeded. Scaling, as discussed in [8], may partially alleviate the problem. Yet the scaling algorithm is complex and does not always work. The authors have seen several well-posed modeling problems involving relatively large populations (e.g., >100) and type D service centers which were not solvable in the range of floating point numbers $1E \pm 75$, despite scaling. The storage requirement is of the order $LK_1K_2\cdots K_R$ as compared to $2K_1K_2\cdots K_R$ for the convolution algorithm.

We now proceed to extend the computational procedure to handle FCFS service centers with multiple constant unit rate servers. The mean value Equation (14) for such a center can be written as

$$T_{r,l} = \frac{\pi_l}{M_l}\left[1 + T_l\left(k-e_r\right) + \sum_{i=0}^{M_l-2}\left(M_l-1-i\right)p_l\left(i,k-e_r\right)\right] \tag{16}$$

where $\left(\pi_l\right)$ is the mean service demand, which is assumed to be independent of the chain. The calculation is complicated by the marginal queue size probabilities, which we have to carry along in the recursive scheme. This can be done by means of Lemma 1, which allows calculation of $p_l\left(i,K\right)$, $i = 1,2,\cdots,M-1$ from previously computed values. In order to keep the recursion going, we need an independent equation for $p_l\left(0,K\right)$, which is obtained from the following relations

$$\sum_{i=0}^{M_l}\left(M_l-i\right)P_l\left(i,k\right) = M_l - \tau_l \tag{17}$$

where

$$\tau_l = \pi_l\sum_{r=1}^{R}\alpha_{r,l} \tag{18}$$

From Equations (17) & (18) we can have the mean number of idle servers as $M_l - \tau_l$, which is just like a little's equation implemented to the set of servers.

b) MULTISERVER CASE
***Step* 1:** Parameters Initialization

$$N_l\left(0\right) = 1,$$
$$P_l\left(0,0\right) = 1,$$
$$P_l\left(l,0\right) = 0, \quad \forall l = 1,2,3,\cdots,L.$$

***Step* 2:** Main Loop, same as in single server case.
***Step* 3:** Additional Corollary for Multi-servers.

$$T'_{r,l} = \frac{p_{r,l}}{M_l} \left\{ N_l\left(i - k_r\right) + \sum_{j=0}^{M_l-2} \left(M_l - 1 - j\right) p_l\left(j, 1 - k_r\right) \right\} \qquad (19)$$

For $r = 1, 2, \cdots, R$ and each FCFS multi server center $l \in S(r)$, while for other service centers use the equation of step 3 in single server case.

Step 4: Little's equation for chains having α_r, same as in single server case.

Step 5: Little's equation for service centers having $N'_l(i)$, same as in single server case.

Step 6: Additional step for calculating marginal queue size under main loop for each multi FCFS service center l and $j = 1, 2, \cdots, M_l - 1$.

$$P_l\left(j, i\right) = \frac{1}{J} \sum_{r \in R(l)} \alpha_r p_{r,l} p_l\left(J - 1, i - e_r\right) \qquad (20)$$

$$\tau_l = \sum_{r \in R(l)} \alpha_r p_{r,l} \qquad (21)$$

$$P_l\left(0, i\right) = 1 - \frac{1}{M_l} \left[\tau_l + \sum_{J=1}^{M_l-1} \left(M_l - j\right) p_l\left(j, i\right) \right] \qquad (22)$$

The process evaluate per multi service center and per recursive step of the order $2(M+1)R$ additions and $3MR + 2M$ multiplications. We observe that it grows linearly with M.

c) Queuing Theory Limitations

The assumptions of classical queuing theory may be too restrictive to be able to model real-world situations exactly. The complexity of production lines with product-specific characteristics cannot be handled with those models. Therefore specialized tools have been developed to simulate, analyze, visualize and optimize time dynamic queuing line behavior.

For example; the mathematical models often assume infinite numbers of customers, infinite queue capacity, or no bounds on inter-arrival or service times, when it is quite apparent that these bounds must exist in reality. Often, although the bounds do exist, they can be safely ignored because the differences between the real-world and theory is not statistically significant, as the probability that such boundary situations might occur is remote compared to the expected normal situation. Furthermore, several studies [9] [10] show the robustness of queuing models outside their assumptions. In other cases the theoretical solution may either prove intractable or insufficiently informative to be useful.

Alternative means of analysis have thus been devised in order to provide some insight into problems that do not fall under the scope of queuing theory, although they are often scenario-specific because they generally consist of computer simulations or analysis of experimental data.

4. Simulation Results & Discussion

The network bottleneck is the fast server. For $\beta > 0.7$ the fast server is also the network bottleneck, but when $\beta < 0.7$, the network bottleneck is the slow server.

The determination of network throughput for different values of β is calculated recursively. Every job arrives at server serve immediately (FCFS). **Figure 2** shows the plot between β and network throughput and clearly shows the difference between two schemes with consistence behavior.

Figure 3 describes value by value β performance with rival scheme. Markov Chain scheme started very confidently achieving 35 Mbps at $\beta = 0.1$ but later at $\beta = 0.5$ our proposed scheme performance increases in terms of Mbps which remain consistent. So in multi-server case for every value of $\beta \geq 0.5$ the queuing systems performs well. After deep analysis on other resulted files created after simulation, we have seen that size of packet is also increases as throughput increase which also help in improving system overall performance.

Figure 4 shows queuing system mean waiting time which increases as usual as the number of customer in chain increases. But still our proposed model is somehow better while comparing with other.

This means waiting time also one of the main objectives of my future work. To define and construct a model in which mean waiting system decreases as the number of customers increases by implementing some grid

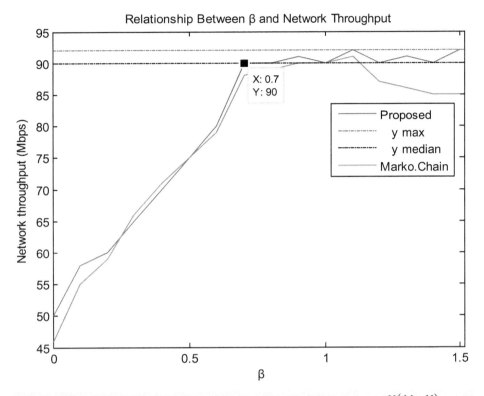

Figure 2. Throughput and consistency behavior between two schemes $\beta_r = \mu \dfrac{N(M-N)}{(M-1)M}$.

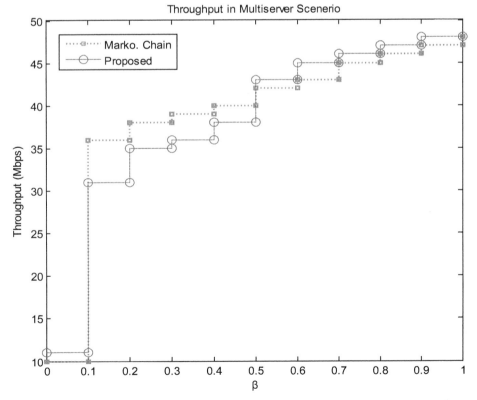

Figure 3. Throughput and consistency behavior (Multi-Server Scenario).

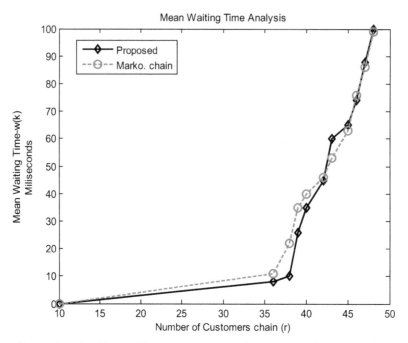

Figure 4. Queuing mean waiting time.

computing functionalities. Also developing a queuing network model for multi-hop wireless ad hoc networks keeping same objectives, used diffusion approximation to evaluate average delay and maximum achievable per-node throughput. Extend analysis to many to one case, taking deterministic routing into account.

Acknowledgements

This research work was partially supported by NSF of china Grant No. 61003247. The authors also would like to thanks the anonymous reviewers and the editors for insightful comments and suggestions.

References

[1] Jones, E.P.C., Li, L. and Ward, P.A.S. (2005) Practical Routing in Delay-Tolerant Networks. *Proceedings of ACM SIGCOMM Workshop on Delay Tolerant Networking (WDTN)*, New York, 1-7.

[2] Zhao, W., Ammar, M. and Zegura, E. (2005) Controlling the Mobility of Multiple Data Transport Ferries in a Delay-Tolerant Network. *INFOCOM*, 1407-1418.

[3] Bambos, N. and Michalidis, G. (2005) Queueing Networks of Random Link Topology: Stationary Dynamics of Maximal Throughput Schedules. *Queueing Systems*, **50**, 5-52. http://dx.doi.org/10.1007/s11134-005-0858-x

[4] Kawanishi, K. (2005) On the Counting Process for a Class of Markovian Arrival Processes with an Application to a Queueing System. *Queueing Systems*, **49**, 93-122. http://dx.doi.org/10.1007/s11134-005-6478-7

[5] Masuyama, H. and Takine, T. (2002) Analysis of an Infinite-Server Queue with Batch Markovian Arrival Streams. *Queueing Systems*, **42**, 269-296. http://dx.doi.org/10.1023/A:1020575915095

[6] Armero, C. and Conesa, D. (2000) Prediction in Markovian Bulk Arrival Queues. *Queueing Systems*, **34**, 327-335. http://dx.doi.org/10.1023/A:1019121506451

[7] Bertsimas, D. and Mourtzinou, G. (1997) Multicalss Queueing System in Heavy Traffic: An Asymptotic Approach Based on Distributional and Conversational Laws. *Operations Research*, **45**, 470-487. http://dx.doi.org/10.1287/opre.45.3.470

[8] Reiser, M. (1977) Numerical Methods in Separable Queuing Networks. *Studies in Management SCI*, **7**, 113-142.

[9] Chandy, K.M., Howard, J., Keller, T.W. and Towsley, D.J. (1973) Local Balance, Robustness, Poisson Departures and the Product Forms in Queueing Networks. Research Notes from Computer Science Department, University of Texas at Austin, Austin.

[10] Knadler Jr., C.E. (1991) The Robustness of Separable Queueing Network Models. In: Nelson, B.L., David Kelton, W. and Clark, G.M., Eds., *Proceedings of the* 1991 *Winter Simulation Conference*, 661-665.

5

An Estimation of Achievable Rate for Digital Transmissions over MIMO Channels

Jinbao Zhang[1,2], Song Chen[1,2], Qing He[1,2]

[1]School of Electronic and Information Engineering, Beijing Jiaotong University, Beijing, China
[2]Beijing Engineering Research Center of EMC and GNSS Technology for Rail Transportation, Beijing, China
Email: jbzhang@bjtu.edu.cn

Abstract

Achievable rate (AR) is significant to communications. As to multi-input multi-output (MIMO) digital transmissions with finite alphabets inputs, which greatly improve the performance of communications, it seems rather difficult to calculate accurate AR. Here we propose an estimation of considerable accuracy and low complexity, based on Euclidean measure matrix for given channel states and constellations. The main contribution is explicit expression, non-constraints to MIMO schemes and channel states and constellations, and controllable estimating gap. Numerical results show that the proposition is able to achieve enough accurate AR computation. In addition the estimating gap given by theoretical deduction is well agreed.

Keywords

Achievable Rate, Digital MIMO Transmissions, Analytical Estimation, Low-Complexity

1. Introduction

Achievable rate (AR), defined as information entropy collected from receiving signals—mutual information, is a fundamental means to evaluate and optimize communications. It is demonstrated that AR is inputs related, and achieves its maximum—channel capacity with Gaussian inputs [1]. Despite optimal, Gaussian inputs are rarely used in practice. Instead, digital transmissions with inputs from finite-alphabet constellations, such as m-PSK and etc., are more common, which depart significantly from Gaussian inputs. Therefore, a considerable AR gap exists between the two inputs [2]. Besides, many results have shown that multi-input multi-output (MIMO) greatly improves the performance of digital transmissions [3]. Consequently AR computation for digital transmissions over MIMO channels is motivated.

Reconsider definition of mutual information in [1]. When it comes to finite-alphabet constellations and

MIMO propagation matrix, it involves multi-dimensional integral to calculate AR, which leads to impractical implementation. And then various estimations are proposed. Monte Carlo method [4] is the most common. Despite accuracy, not only it is too implicit for analytical applications, but also it costs too much computational complexity as order of modulation and MIMO increases. Particle method [5] is proposed to reduce the complexity. However, it remains implicit. Aiming analytical solution, lower bounds and approximations for AR are proposed in [2] [6] and [7] respectively, showing validity under certain scenarios. Unfortunately, limitation remains. Lower bound in [2] requires unitary inputs—matrix having orthonormal columns [8]. For wireless MIMO channels, such assumption is rarely achieved. As to approximations in [6] [7], constellations and MIMO channels related tuning factor is indispensable to the estimating accuracy, which also introduces limitation. Moreover, gap between true AR and lower bounds/approximations is not analyzed in the mentioned work.

Comparing to current proposals, the main contribution of this letter is to propose AR estimation of low complexity, analytically explicit expression and controllable gap, without constraints to inputs and MIMO channels. This work is organized as follows. Section 2 formulates the problem and premiss; Section 3 describes details of proposed solution, and analyzes estimating gap; Section 4 gives numerical results, and further discuss on computational complexity and estimating gap; finally conclusions are drawn in Section 5.

2. Problem Formulation

2.1. Notations and Definitions

This work uses the following notations. Italic character in lower and upper case denotes variable. Bold italic character in lower and upper case denotes vector and matrix respectively. The superscript $(\bullet)^H$ denotes conjugate transposition. $\boldsymbol{I}_N$ is $N \times N$ identity matrix. $[A]_{k,m}$ is the element of matrix A at k^{th} row and m^{th} column. $[\boldsymbol{a}]_k$ is the kth element of vector $\boldsymbol{a}$. $\text{tr}(\bullet)$ denotes the trace of square matrix. $\|\bullet\|$ is Euclidean norm of matrix and vector. $E\{\bullet\}$ denotes the expectation of random variable. $\delta(x)$ is Dirac delta function. $\mathcal{CN}(0,\sigma^2)$ is a complex Gaussian random scalar, and its real and imaginary components are independent and identically normal distributed with zero-mean and variance of $\sigma^2/2$. $\mathcal{N}(0,\sigma^2)$ is a real Gaussian random scalar normal distributed with zero-mean and variance of σ^2. $\mathbb{C}$ is the complex space, and $\mathbb{R}$ is the real space. $\mathbb{A}=\{a_k\}$ means that space—$\mathbb{A}$ is consisted of elements—a_k. $\mathbb{A}^{a\times b}$ denotes $a\times b$ tensor space based on $\mathbb{A}$. Operation $\otimes$ is cartesian product of space.

2.2. Signal Model and Premise

Consider digital transmissions over $N_R \times N_T$ MIMO channels, following assumptions are premised.

- $\boldsymbol{H} \in \mathbb{C}^{N_R \times N_T}$, is an $N_R \times N_T$ complex propagation matrix, known to receiver.
- $\boldsymbol{x} \in \mathbb{C}^{N_T \times 1}$, is an $N_T \times 1$ transmitted symbol vector. $[\boldsymbol{x}]_k$ is independently and uniformly selected from k^{th} normalized finite-alphabet constellation—$\mathbb{Q}(k)=\{q_{m,k}\}$: $m=1,2,\cdots,N_k$. Moreover, $\mathbb{Q}(k)$ may differ with k. Thus

$$p([\boldsymbol{x}]_k) = \frac{1}{N_k} \sum_{m=1}^{N_k} \delta([\boldsymbol{x}]_k - q_{m,k}). \tag{1}$$

Define $\mathbb{Q}=\mathbb{Q}(1)\otimes\mathbb{Q}(2)\otimes\cdots\otimes\mathbb{Q}(N_T)$. Consequently $\boldsymbol{x} \in \mathbb{Q}=\{\boldsymbol{q}_k\}$: $k=1,2,\cdots,N$, and $N=\Pi_{k=1}^{N_T}$. So we have

$$\underset{\boldsymbol{x}\in\mathbb{Q}}{E}\{\boldsymbol{x}\boldsymbol{x}^H\} = \boldsymbol{I}_{N_T}, \quad p(\boldsymbol{x}) = \frac{1}{N}\sum_{k=1}^{N}\delta(\boldsymbol{x}-\boldsymbol{q}_k). \tag{2}$$

- $\boldsymbol{w} \in \mathbb{C}^{N_R \times 1}$, is an $N_R \times 1$ independent complex additional white Gaussian noise (AWGN) vector, and

$$[\boldsymbol{w}]_k = \mathcal{CN}(0,\sigma^2), \quad \underset{\boldsymbol{w}\in\mathbb{C}^{N_R\times1}}{E}\{\boldsymbol{w}\boldsymbol{w}^H\} = \sigma^2 \boldsymbol{I}_{N_R}. \tag{3}$$

- $\boldsymbol{y} \in \mathbb{C}^{N_R \times 1}$, is an $N_R \times 1$ receiving symbol vector. Hence, AR is defined as mutual information [1],

$$I = \underset{\boldsymbol{x}\in\mathbb{Q};\boldsymbol{y}\in\mathbb{C}^{N_R\times1}}{E}\left\{\lg_2 \frac{p(\boldsymbol{x},\boldsymbol{y})}{p(\boldsymbol{x})p(\boldsymbol{y})}\right\}. \tag{4}$$

Note that, Equation (4) is AR value for the whole transmitting and receiving vector. Considering spatial multiplexing mode of MIMO, AR value for each element in x is also needed, so we formulate the problem as estimation of AR for both vector and each element in the vector.

3. Low-Complexity Solution

Firstly, Equation (4) is rewritten as

$$I = \int_{x \in \mathbb{Q}; y \in \mathbb{C}^{N_R \times 1}} p(x)p(y|x)\lg_2 \frac{p(y|x)}{\int_{\mathbb{Q}} p(x)p(y|x)dx} dxdy. \tag{5}$$

Given H and σ^2, the posterior probability is,

$$p(y|x) = \frac{1}{N_R \sigma^{2N_R}} e^{-\frac{\|y-x\|^2}{\sigma^2}}. $$

Given $\mathbb{Q}$, we have

$$I = \frac{1}{N} \sum_{k=1}^{N} \int_{\mathbb{C}^{N_R \times 1}} e^{-\frac{\|w\|^2}{\sigma^2}} \lg_2 \left(\frac{e^{-\frac{\|w\|^2}{\sigma^2}}}{\frac{1}{N} \sum_{m=1}^{N} e^{-\frac{\|H(q_k-q_m)+w\|^2}{\sigma^2}}} \right) dw. \tag{6}$$

Definition: Euclidean measure matrix D for given constellation $\mathbb{Q}$ and H is as

$$[D]_{k,m} = \|H(q_k - q_m)\|. \tag{7}$$

Recall Equation (3), and then Equation (6) is rewritten as

$$I = \frac{1}{N} \sum_{k=1}^{N} \underset{w'_{k,m} \in \mathbb{R}}{E} \left\{ \lg_2 \left(\frac{1}{N} \sum_{m=1}^{N} e^{-\frac{[D]_{k,m}^2 + w'_{k,m}}{\sigma^2}} \right) \right\}, \tag{8}$$

where $w'_{k,m}$ is combination of AWGNs,

$$w'_{k,m} = \operatorname{tr}\left\{ H(q_k - q_m)w^H + w(q_k - q_m)^H H^H \right\}. \tag{9}$$

Then normalize $w'_{k,m}$ as

$$w = w'_{k,m} / \left(\sqrt{2}[D]_{k,m}\sigma \right) = \mathcal{N}(0,1). \tag{10}$$

So we rewrite Equation (8) as

$$I = \frac{1}{N} \sum_{k=1}^{N} \int_{\mathbb{R}} \frac{e^{-\frac{w^2}{2}}}{\sqrt{2\pi}} \lg_2 \left(\frac{1}{N} \sum_{m=1}^{N} e^{-\frac{[D]_{k,m}^2}{\sigma^2} - \frac{\sqrt{2}[D]_{k,m}}{\sigma}w} \right) dw. \tag{11}$$

Theorem 1. *AR is estimated by exponentially weighted average of Euclidean measure matrix as*

$$I \approx \tilde{I} = -\frac{1}{N} \sum_{k=1}^{N} \lg_2 \left(\frac{1}{N} \sum_{m=1}^{N} e^{-\frac{[D]_{k,m}^2/\sigma^2}{3 - e^{-[D]_{k,m}^2/(4\sigma^2)}}} \right) \tag{12}$$

Proof. see **Appendix 1**.
Theorem 2. *AR is decomposed as*

$$I\left(x_k; y\right) = I - I\left(x_k; y\right),\qquad(13)$$

where x_k denotes the sub-vector excluding x_k from x.

Proof. see **Appendix 2**.

With Theorem 1 and 2, the formulated problem in Section 2 is solved.

Theorem 3. *Gap between true and estimated AR given in proposition 1 is bounded by exponentially weighted average of minimum Euclidean measure as*

$$\left| I - \tilde{I}\right| \leq \left| e^{-\frac{\hat{d}_k^2}{2\sigma^2}} - \frac{1}{N}\sum_{m=1}^{N} e^{-\frac{\hat{d}_k^2/\sigma^2}{3 - e^{-\hat{d}_k^2/\left(4\sigma^2\right)}}}\right|,\qquad(14)$$

where $\hat{d}_k$ denotes the minimum among non-zero elements in k^{th} column of Euclidean measure matrix.

Proof. see **Appendix 3**.

Recalling Theorem 3, the gap between true and estimated AR for x_k is

$$\left| I\left(x_k; y\right) - \tilde{I}\left(x_k; y\right)\right| \leq \left| I - \tilde{I}\right| + \left| I\left(x_k; y\right) - \tilde{I}\left(x_k; y\right)\right|.\qquad(15)$$

4. Numerical Results

To verify Theorem 1, 2 and 3, numerical results are provided. For generality, 2×3 complex propagation matrixes consisted of independent Gaussian distributed elements are used as MIMO channels,

$$H \in \mathbb{C}^{2\times3}, \quad \left[H\right]_{k,m} = \mathcal{CN}\left(0,1\right).\qquad(16)$$

Also high and low correlated scenarios with correlation coefficient of 0.1 and 0.9 are considered respectively,

$$H_{\text{corr}} = R_R^{1/2} H R_T^{1/2}.\qquad(17)$$

where correlation matrices R_R and R_T are generated with spatial channel model and correlation coefficient [3].

The 3 transmitted symbols in x are modulated by BPSK, 8PSK and 16QAM respectively. True AR is computed by Monte Carlo method [4]. And estimated AR is computed with Theorem 1 and 2.

Numerical results in **Figure 1** and **Figure 2** show that the proposed estimation is able to achieve enough accurate AR over complex Gaussian random MIMO channel. And according to Equation (6) and (12), the calculation complexity is reduced to $2N^2$ exponentiations and N logarithms, instead of N integrals.

Despite slightness, numerical results show that the gap remains. However, such estimating gap can be quantized by Equation (14) and (15), and numerical results are shown in **Figure 3**. According to **Figure 1** and

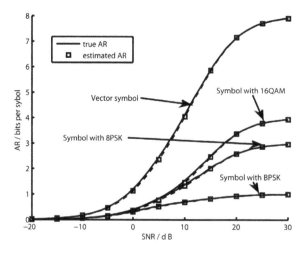

Figure 1. AR for 2 × 3 random Gaussian distributed MIMO channels with transmitting and receiving correlation coefficient of 0.1.

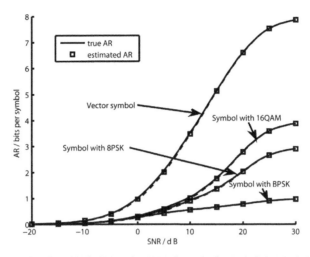

Figure 2. AR for 2 × 3 random Gaussian distributed MIMO channels with transmitting and receiving correlation coefficient of 0.9.

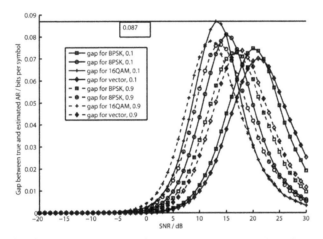

Figure 3. Gap between true and estimated AR for 2 × 3 random Gaussian distributed MIMO channels with transmitting and receiving correlation coefficient of 0.1 and 0.9, computed with Equations (14) and (15).

Figure 2, the maximum gap between estimated and true AR is lower than 0.0733 bits/symbol, which agrees well to theoretic bound of 0.0603 bits/symbol given in **Figure 3**, which is computed by Theorem 3.

5. Conclusion

A low-complexity AR estimation is presented in this work. Numerical results show that it is accurate enough, and the deductive theoretic bound of estimating gap is well matched. Moreover, the most encouraging thing is that, the proposed estimation is of no constraints to finite-alphabet constellations and MIMO channels. Besides, as shown in Equation (12), this proposition deduces integral of AR calculation into an weighted average of Euclidean measure matrix for given channel states and constellations, which is explicit enough for analytical applications.

Acknowledgements

We thank the Editor and the referee for their comments. This work is funded by the NSFC program (61172021 and 61471030), the Fundamental Research Funds for the Central Universities (2014JBZ023), Beijing city science and technology special program (Z141101004414091), and Research on the development of science and technology plan Chinese Railway Corporation (2014X012-B, Z2014-X002). This support is greatly appreciated.

References

[1] Shannon, C.E. (1948) A Mathematical Theory of Communication. *The Bell System Technical Journal*, **27**, 379-423. http://dx.doi.org/10.1002/j.1538-7305.1948.tb01338.x

[2] Zeng, W.L., Xiao, C.S., Wang, M.X. and Lu, J.H. (2012) Linear Precoding for Finite-Alphabet Inputs Over MIMO Fading Channels with Statistical CSI. *IEEE Transactions on Signal Processing*, **60**, 3134-3148. http://dx.doi.org/10.1109/TSP.2012.2188717

[3] Mesleh, R.Y., Haas, H., Sinanovic, S., Ahn, C.W. and Yun, S. (2008) Spatial Modulation. *IEEE Transactions on Vehicular Technology*, **57**, 2228-2241. http://dx.doi.org/10.1109/TVT.2007.912136

[4] Greenspan, D. and Casulli, V. (1994) Numerical Analysis for Applied Mathematics, Science, and Engineering. Westview Press, Boulder.

[5] Dauwels, J. and Loeliger, H.A. (2008) Computation of Information Rates by Particle Methods. *IEEE Transactions on Information Theory*, **54**, 406-409. http://dx.doi.org/10.1109/TIT.2007.911181

[6] Alvarado, A., Brannstrom, F. and Agrell, E. (2014) A Simple Approximation for the Bit-Interleaved Coded Modulation Capacity. *IEEE Communications Letters*, **18**, 495-498.

[7] Zhang, J.B., Zheng, H.M., Tan, Z.H., Chen, Y.Y. and Xiong, L. (2010) Link Evaluation for MIMO-OFDM System with ML Detection. 2010 *IEEE International Conference on Communications*, Cape Town, 23-27 May 2010, 1-6. http://dx.doi.org/10.1109/ICC.2010.5502079

[8] Hassibi, B. and Marzetta, T.L. (2002) Multiple-Antennas and Isotropically Random Unitary Inputs: The Received Signal Density in Closed Form. *IEEE Transactions on Information Theory*, **48**, 1473-1484. http://dx.doi.org/10.1109/TIT.2002.1003835

Appendix

1. Proof of Theorem 1

Proof. Following approximations are easily achieved with numerical methods,

$$\forall x \in [0,1], \quad \exists \beta = 0.38, \text{ s.t. } \max \left\{ \left| \lg_e (1+x) - x e^{-\beta x} \right| \right\} \approx 0,$$

$$\forall x > 0, \quad \exists \gamma = 0.68, \text{ s.t. } \max \left\{ \left| Q(x) - e^{-x^2/2 - \gamma x} / 2 \right| \right\} \approx 0. \tag{18}$$

Although gap still remains, following deduction will show that such gap makes no difference on AR computations. Define

$$\rho_{k,m} = [\boldsymbol{D}]_{k,m}^2 / \sigma^2. \tag{19}$$

To prove of Equation (12) equals to prove

$$\mathcal{F}(\boldsymbol{\rho}_{k,N}) = \int_{\mathbb{R}} \frac{e^{-\frac{w^2}{2}}}{\sqrt{2\pi}} \lg_2 \left(\sum_{m=1}^{N} e^{-\rho_{k,m} - \sqrt{2\rho_{k,m}} w} \right) dw, \tag{20}$$

where $\boldsymbol{\rho}_{k,N} = [\rho_{k,1}, \rho_{k,2}, \cdots, \rho_{k,N}]$ is an $N \times 1$ ascending sorted vector, and

$$\mathcal{F}(\boldsymbol{\rho}_{k,N}) \approx \hat{\rho}_k = -\lg_2 \left(\sum_{m=1}^{N} e^{-\frac{\rho_{k,m}}{3 - e^{-\rho_{k,m}/4}}} \right). \tag{21}$$

Use inductive reasoning, define $g_{k,N}$ as the gap,

$$\mathcal{F}\left(\rho_{k,N}\right) = \hat{\rho}_{k,N} + g_{k,N}.\tag{22}$$

For $N = 1$, recalling Equation (7), $\rho_{k,1}$ constantly equals to 0,

$$g_{k,1} = \frac{\rho_{k,1}}{3 - e^{-\rho_{k,1}/4}} - \rho_{k,1} \rightsquigarrow \mathcal{F}\left(\rho_{k,1}\right) = \hat{\rho}_{k,1}.\tag{23}$$

Then assume

$$g_{k,M} \approx 0 \rightsquigarrow \mathcal{F}\left(\rho_{k,M}\right) = \hat{\rho}_{k,M}.\tag{24}$$

This implies that, within the operative domain of

$$\int_{\mathbb{R}} \frac{e^{-\frac{w^2}{2}}}{\sqrt{2\pi}} \lg_2(\bullet)\,dw,$$

following approximation is valid.

$$\sum_{m=1}^{M} e^{-\rho_{k,m} - \sqrt{2\rho_{k,m}}\,w} \approx e^{-\hat{\rho}_{k,M} - \sqrt{2\hat{\rho}_{k,M}}\,w}\tag{25}$$

Recall Equation (18), for $M + 1$, we have

$$\begin{aligned}
\mathcal{F}\left(\rho_{k,M+1}\right) &= \int_{\mathbb{R}} \frac{e^{-\frac{w^2}{2}}}{\sqrt{2\pi}} \lg_2\left(\sum_{m=1}^{M+1} e^{-\rho_{k,m} - \sqrt{2\rho_{k,m}}\,w}\right)dw \\
&\approx \int_{\mathbb{R}} \frac{e^{-\frac{w^2}{2}}}{\sqrt{2\pi}} \lg_2\left(e^{-\hat{\rho}_{k,M} - \sqrt{2\hat{\rho}_{k,M}}\,w} + e^{-\rho_{k,M+1} - \sqrt{2\rho_{k,M+1}}\,w}\right)dw \\
&= \hat{\rho}_{k,M} - \Delta_{k,M+1},
\end{aligned}\tag{26}$$

where

$$\begin{aligned}
\Delta_{k,M+1} &\approx e^{-\frac{\left(\sqrt{\hat{\rho}_{k,M}} + \sqrt{\rho_{k,M+1}}\right)^2}{4}} \left[\left(\hat{\rho}_{k,M} - \rho_{k,M+1}\right) + \left(\sqrt{\hat{\rho}_{k,M}} - \sqrt{\rho_{k,M+1}}\right)\Big/\sqrt{\pi} - \left(\hat{\rho}_{k,M}/2 - \rho_{k,M+1}/2\right)e^{-\gamma\left(\sqrt{\hat{\rho}_{k,M}/2} + \sqrt{\rho_{k,M+1}/2}\right)}\right. \\
&\quad \left. + \frac{1}{2}e^{\gamma\left(3\sqrt{\hat{\rho}_{k,M}/2} - \sqrt{\rho_{k,M+1}/2}\right) - \beta e^{\gamma}\left(\sqrt{2\hat{\rho}_{k,M}} - \sqrt{2\rho_{k,M+1}}\right)}\right].
\end{aligned}\tag{27}$$

And the right side of Equation (22) is

$$\begin{aligned}
\hat{\rho}_{k,M+1} &= -\lg_e\left(\sum_{m=1}^{M} e^{-\frac{\rho_{k,m}}{3 - e^{-\rho_{k,m}/4}}} + e^{-\frac{\rho_{k,M+1}}{3 - e^{-\rho_{k,M+1}/4}}}\right) \\
&\approx \hat{\rho}_{k,M} - e^{-\hat{\rho}_{k,M} - \frac{\rho_{k,M+1}}{3 - e^{-\rho_{k,M+1}/4}}} e^{-\beta e^{-\hat{\rho}_{k,M} - \frac{\rho_{k,M+1}}{3 - e^{-\rho_{k,M+1}/4}}}}.
\end{aligned}\tag{28}$$

So that

$$g_{k,M+1} \approx e^{-\hat{\rho}_{k,M} - \frac{\rho_{k,M+1}}{3 - e^{-\rho_{k,M+1}/4}}} e^{-\beta e^{-\hat{\rho}_{k,M} - \frac{\rho_{k,M+1}}{3 - e^{-\rho_{k,M+1}/4}}}} - \Delta_{k,M+1}.\tag{29}$$

Using monotonic property of exponentiation and logarithm, it is demonstrated that,

$$0 \le g_{k,M+1} \le e^{-\hat{\rho}_{k,M} - \frac{\rho_{k,M+1}}{3 - e^{-\rho_{k,M+1}/4}}} \le e^{-\hat{\rho}_{k,M-1} - \frac{\rho_{k,M}}{3 - e^{-\rho_{k,M}/4}}} \le g_{k,M} \approx 0\tag{30}$$

Recalling Equation (23), (24) and (30), assigning $M = N$, Equation (21) is proved. And then Theorem 1 is proved by substituting Equation (21) in Equation (11).

2. Proof of Theorem 2

Proof. Use $\mathbb{Q}_s$ and $\mathbb{Q}_r$ to denote the sub-set of all possible values of x_s and x_r, and then recall Equation (5), using x_s to denote the targeting sub-vector of computation, and x_r to denote the sub-vector excluding x_s from x. We have,

$$
\begin{aligned}
I(x_s;y) &= \int_{x\in\mathbb{Q}_s;\,y\in\mathbb{C}^{N_R\times 1}} p(x)p(y|x)\lg_2\frac{p(y|x)}{\int_\mathbb{Q}p(x)p(y|x)\mathrm{d}x}\mathrm{d}x\mathrm{d}y \\
&= \int_{x\in(\mathbb{Q}-\mathbb{Q}_s);\,y\in\mathbb{C}^{N_R\times 1}} p(x)p(y|x)\lg_2\frac{p(y|x)}{\int_\mathbb{Q}p(x)p(y|x)\mathrm{d}x}\mathrm{d}x\mathrm{d}y \\
&= \int_{x\in\mathbb{Q};\,y\in\mathbb{C}^{N_R\times 1}} p(x)p(y|x)\lg_2\frac{p(y|x)}{\int_\mathbb{Q}p(x)p(y|x)\mathrm{d}x}\mathrm{d}x\mathrm{d}y \\
&\quad - \int_{x\in\mathbb{Q}_r;\,y\in\mathbb{C}^{N_R\times 1}} p(x)p(y|x)\lg_2\frac{p(y|x)}{\int_\mathbb{Q}p(x)p(y|x)\mathrm{d}x}\mathrm{d}x\mathrm{d}y \\
&= I - I(x_r;y).
\end{aligned}
\tag{31}
$$

Then designate x_k to denote the sub-vector excluding x_k from x, and Theorem 2 is proved.

3. Proof of Theorem 3

Proof. Recall Equation (12) and (30), the maximum gap between true and estimated AR value for x is as follows,

$$
\left|I-\tilde{I}\right| \le \frac{1}{N^2}\sum_{k=1}^{N}\sum_{m=1}^{N}g_{k,m} \le \frac{1}{N}\sum_{k=1}^{N}g_{k,2} \le \frac{1}{N}\sum_{k=1}^{N}\left|e^{-\frac{\rho_{k,2}}{3-e^{-\rho_{k,2}/4}}}-e^{-\frac{\rho_{k,2}}{2}}\right|.
\tag{32}
$$

Since that the sequence of ρ_k is ascending sorted, and then we have,

$$
\rho_{k,2} = \min_{m\in\{2,3,\cdots,N\}}\left\{\frac{[D]_{k,m}^2}{\sigma^2}\right\} \overset{\Delta}{=} \hat{d}_k.
\tag{33}
$$

Theorem 3 is proved.

6

FMAC: Fair Mac Protocol for Achieving Proportional Fairness in Multi-Rate WSNs

Nusaibah M. Al-Ratta, Mznah Al-Rodhaan, Abdullah Al-Dhelaan

Computer Science Department, College of Computer and Information Sciences, King Saud University, Saudi Arabia
Email: nsoobyal-ratta@hotmail.com, rodhaan@ksu.edu.sa, rodhaan@ksu.edu.sa

Abstract

In a multi-rate wireless environment, slow nodes occupy the channel for longer time than fast nodes and thus the total throughput of the network will be reduced. In this research, we study the problem of fairness in multi-rate wireless sensor networks. To improve the fairness, we propose a new protocol, FMAC (Fair MAC protocol) that is based on IEEE 802.11 MAC protocol to achieve proportional fairness between all nodes. FMAC protocol includes medium delay periods within Backoff algorithm to utilize the idle slots of time and reduce the number of collisions and then number of retransmissions, and thus reducing the energy consumption, which is very critical in wireless sensor networks. The experimental results show that transmissions become faster with less collisions and power consumption when applying FMAC, while the aggregated throughput and proportional fairness are increased. The detailed performance evaluation and comparisons are provided using the simulation.

Keywords

Backoff, DCF, IEEE 802.11, Proportional Fairness, Wireless Sensor Networks

1. Introduction

Wireless Sensor Network (WSN) nowadays is regarded as one of the most recent technologies. It is an emerging and fast progressing field in the 21st Century. It has appeared as a result of increasing requirements of mobility in the world, while the traditional networks have proven that they cannot face the challenges of our new life-styles.

Wireless sensor networks were developed initially to be used in the military field like enemy monitoring and battlefield reconnaissance [1]. Today, these networks have a variety of applications especially for tracking outdoor environments, which requires long-range operation for long times with high accuracy and with less maintenance [2]. They are, in essence, a type of Ad-Hoc network in which sensor nodes (SNs) are connected together with wireless links and can join or disconnect from the network with minimal cost and effort.

Wireless sensor networks (WSNs) comprise many of wireless sensor nodes that may be organized in multi-hop or one-hop network with varying quality of wireless link, especially in the case of network with a high traffic, as well as various distances from the sink node. In order to get accurate information from the sensing application, we have to insure the successful fair rate for data transferring from the sensors to the sink [3] [4]. Providing the maximum throughput with minimum power consumption is one of the greatest goals in WSNs. MAC layer is an important place to make the significance improvements towards achieving such goals, due to its responsibility for distribution of the shared resources and controlling the access of the contending nodes to the wireless channel. An efficient MAC protocol can play a big role in power saving, especially regarding the reduction of collisions and thus retransmissions which results in high power consumption [5]-[7].

Collision is not only one of the most energy consumption sources; it also increases latency. So, reducing the collisions can improve the throughput. An efficient MAC protocol can contribute in prolonging the sensor lifetime [5] [8].

The use of IEEE 802.11 protocol in WSNs is considered a new orientation to some extent. The efficiency of the IEEE 802.11 MAC protocol in WSNs has been proven in terms of many important issues like the throughput, delay and power consumption. Because of this and since we know that the channel access mechanism has a large effect on the performance of any network, we decided to design a light and efficient solution which is based on IEEE 802.11 MAC protocol to achieve the proportional fairness in multi-rate WSNs.

Our main objective is to achieve the proportional fairness between all nodes in multi-rate WSNs via designing a new MAC protocol. This new protocol will be able to investigate the trade-off among throughput and fairness.

We also aim to make an in-depth analysis and validation of the performance of the proposed protocol which can help in advancing research about the fairness problem using the distributed coordination function (DCF) for the IEEE 802.11 wireless sensor networks.

Many research papers related to the fairness issue were suggested solutions that include modifications of the size of contention window or even the size of the packets in order to restrict the transmission of some specific nodes, or adding some control information which may result in additional overhead or energy wastage. Also some solutions need a supplementary alteration for the upper layers characteristics such as routing issues.

Our proposed solution is lightweight in terms of not requiring any additional information or lot of computations. It includes a small modification in the standard backoff process to give different opportunities to the nodes that have different data rates. This work will hopefully promote the scientific and technological research in wireless sensor networks. However, the proposed protocol is also applicable in WLANs because it does not mainly depend on the nature of WSNs.

The rest of this paper is organized as follows: Section 2 defines the problem statement. Section 3 provides necessary background information. Section 4 gives a rich review of the related work. Then we present the methodology in Section 5. Next, the simulation and performance evaluation metrics as well as the experimental results are provided in detail in Section 6 and 7 respectively. Finally Section 8 concludes the paper and provides some future directions.

2. Problem Statement

The IEEE 802.11 uses the Distributed Coordination Function (DCF) to regulate transmissions at media access control (MAC) layer [9]. DCF provides long-term equal transmission opportunities to all competing nodes regardless of their data rates, which results in throughput-based fairness and therefore significantly degrades the aggregated throughput in a multi-rate wireless sensor network (WSN). This is because of the nodes with low-rates that need longer occupancy time of the channel for transmission, which will result in shortening the time that the fast nodes can find the channel as idle.

This performance anomaly can be mitigated by adjusting the size of contention window, increasing contention window size for a specific node will reduce its competition opportunity for example, or adopting a more appropriate backoff algorithm [9]. On the other hand, proportional fairness intends to maximize the sum of logarithm of the throughput for all nodes while providing fairness in terms of throughput. Based on these observations, we propose a new MAC protocol for multi-rate WSN with a single sink node to achieve proportional fairness.

3. Background Information

In this Section, we will provide some necessary background information about the MAC layer protocols for WSN.

3.1. MAC Protocols for WSNs

In the traditional IEEE 802 local area networks (LANs), Data Link Layer is split into two sub layers: Logical Link Control (LLC) and Media Access Control (MAC). A new MAC sub layer for 802.11 was then developed. The main functions of MAC layer involve frame delineation, addressing, error checking, and organize access of all nodes to the shared-medium [10].

The protocols for MAC layer in WSNs should be energy efficient in order to prolong lifespan of the nodes. The main objectives of such protocols include maximizing channel capacity usage while minimizing the delay time. Other goals involve fairness and stability [11].

MAC is a fundamental method to control the successful transfer processes in the network. We can group existing MAC protocols for WSN in two main categories as follows: contention-based category and contention-free category [12].

Contention-free protocols have a fixed assignment such as CDMA, TDMA and FDMA. Lack of flexibility in resource allocation is the main drawback of these protocols. So, they faced difficulties with changes in the configuration, and thus they are inappropriate in the dynamic wireless networks.

On the other hand, contention-based protocols have random assignment like CSMA and ALOHA. Such protocols have a high flexibility and they are widely applied in the wireless LANs. Systems can be developed by the use of more than one of these categories [11].

Most of the suggested protocols in the category of contention-based are used the techniques of carrier sense multiple access (CSMA). In CSMA, the node has special carrier sensing capabilities. Before starting transmitting, the node must sense the channel. If it found that the channel is busy, then the access will be postponed to the next retrying [12].

CSMA has two extensions, collision detection (CSMA/CD) and collision avoidance (CSMA/CA) [11]. In collision detection, the sender will be able to detect the collision after it happens and thus stop transmitting data. In collision avoidance, the node applies technique that avoids the collision before it happens, but with no guarantees.

Usually it is difficult to detect collisions in a wireless node. Many methods were proposed to address this problem, but even these approaches unable to grants enough ability for nodes to detect all collisions types. Therefore, IEEE 802.11 MAC protocol with CSMA/CA was suggested to be the standard protocol used in wireless local area networks (LANs) [11] [13].

Two access methods are provided by IEEE 802.11 access scheme: Distributed Coordination Function (DCF) and Point Coordination Function (PCF). The former is for contention-based access scheme with asynchronous transferring, while the latter is for the contention-free and centralized access scheme [14]. The DCF method is involved in the CSMA/CA category for MAC protocols.

3.2. Binary Exponential Backoff

Wastage of channel utilization is mainly caused by collisions as well as the idle times resulted by the access distribution [14]. Binary exponential backoff (BEB) is regarded as an adaptive behavior provided by the IEEE 802.11 DCF mechanism in order to address such wastage.

Each node wants to send a frame, sets a random time for accessing the channel, this amount of time is varying based on the number of collisions which occurred earlier for that frame. If an ACK frame was not received, a node will assume that the corresponding packet was dropped by the collision, and thus it will retransmit this packet after invoking BEB. Backoff time is increased using BEB technique at each time the collision occurs, because this may indicate that the network is congested.

The BEB [9] [12]-[14] works as follows: Slot_Time is a unit of a constant length used to measure the time, also can be just called slot. In order to implement Binary Exponential Backoff scheme, each node has the Backoff Counter (BC), which is the counter used to compute the required empty slots of time for waiting before attempting to transmit. Before scheduling of a new transmission, the node should wait for random interval which is within a range that is uniformly distributed between 0 and the size of the current contention window (CW).

Backoff value is defined as the following:

$$BC = Random(\ )*CW\_Size. \tag{1}$$

where the function Random ( ) returns a number selected randomly between 0 and 1.

If the node senses that the channel is busy, it will pause its backoff counter and postpone the transmission. After the end of the ongoing transmission, if it senses that the channel is idle then it will resume the counter again. At each time slot, the value of backoff counter is reduced by one. If its value becomes zero, and if the channel is found not busy for the time interval Distributed Inter Frame Space (DIFS), the node will (re)start trying to access it again.

If the node accessed the channel and successfully sent the packet, it will wait for the acknowledgment (ACK) frame from the receiver within SIFS interval (Short Inter Frame Space), which is less than DIFS. On the other hand, if the node needs to access but the channel is busy, it will stop the decreasing of the counter and preserve its value to be resumed in the next attempt of transmission.

If two nodes chose the same random backoff value, both of them will start transmission simultaneously, and thus the collision will occur. In this case, the sender nodes will detect the collision because of the absence of an ACK for their frames. So, every node will apply the exponential backoff algorithm. Firstly, the size of the contention window (CW) will be doubled. Next, it will choose a random number and assign it to the value of its backoff counter. Then, it will start decreasing the counter until it equals zero. Finally, the node will attempt to retransmit again. If it is successful, it will reset the value of CW to the minimum CWmin.

The initial value of contention window size is CWmin. At each unsuccessful (re)transmission, the size of contention window will be doubled, to reduce the contention. If this size reaches the maximum value CWmax, the protocol will stop the retransmission, discard all not transmitted packets, and reset the size to CWmin. **Figure 1** shows the DCF protocol [15].

3.3. DCF and RTS/CTS Access Mechanism

RTS/CTS exchanging can optionally extend the Binary exponential Backoff access mechanism of IEEE 802.11 DCF. When the node gets to access the channel, it will send the announcement for its incoming transmission rather than sending the actual data packets. This can be applied by sending a request to send (RTS) control packet to the receiver.

If the receiver is ready to receive the data, it will reply by sending a clear to send (CTS) control packet. Using this technique, all other nodes within the range will be informed about this transmission. So, they will not disturb it. Both the control packets include the expected transmission length, thus the channel will be reserved efficiently for this transmission [16].

3.4. Performance Anomaly

In wireless channels, there are many reasons that affect the quality of signals, such as the noise, attenuation and interference. This may cause unsuccessful reception of the frames. The implications of these errors are more

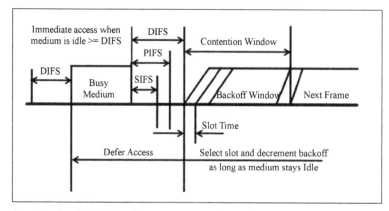

Figure 1. DCF protocol.

critical with 802.11 DCF, because it is not possible to distinguish between fail transmissions and collisions. When the frame is lost, the node will implement the exponential backoff process.

If the node uses a lower bit rate, this may result in a lower frame error rate and thus a better throughput. Unfortunately, diversity of bit rates, either by changing the bit rate for the same node or by the existence of nodes with different bit rates originally, causes a performance anomaly. Slower rate nodes consume longer time for transmission than faster rate nodes. As a result, there will be less time for fast nodes with an idle channel, and thus the throughput will be reduced [9].

4. Related Work

Congestion control has a significant importance in wireless sensor networks. Scaling up wireless sensor networks by adding more sensors in a larger area implies increasing the traffic volume while the capacity of the channel around the bottlenecks cannot be increased easily, specifically with low-cost and low-energy constraints.

Resolving congestion does not guarantee fairness [17], because in reality they are two different problems, although they are related in terms of their effects on the throughput of the network.

Fairness is one of the most important goals in wireless sensor networks. This issue becomes even more serious for networks with multi data rates. Usually, nodes with the low data rates have a longer occupancy time of the channel than other nodes, and thus will degrade the whole throughput of the network. Nodes may use different data rates because of the conditions of the channel, different generations of technologies, or simply based on their distance from the sink node [18].

As wireless sensor networks are usually used in surveillance fields, the fair access to the network between all nodes must be ensured in order to enable the sink nodes to get a complete view of the monitored area. Cooperative MAC Protocol suggested in [19] to tackle this problem. Low data rate node assisted by a high rate node (called helper node) in its transmission. In order to reduce the occupancy time of the channel, it used two transmissions with high data rates, rather than only one transmission with a low data rate.

Fairness has different definitions based on different criteria. Regarding time scales, there are two types of fairness: the first type is called long-term fairness while the second is called short-term fairness [9].

Long-term fairness gives equal probabilities of accessing the channel successfully among all competing nodes on a long-term scale. Short-term fairness gives the same thing, but within shorter time intervals. With short-term fairness, each node can access the channel after a short period of time that results in brief delays. Actually, short term fairness fulfils the long-term fairness, but not vice versa.

IEEE 802.11 MAC protocol provides good short-term fairness for the networks with a limited number of nodes [9]. But with large number of nodes, it does not provide short term fairness due to the use of exponential backoff by nodes that have collided packets, and thus results in more opportunities of transmission for other nodes.

All nodes choose their random backoff intervals from the similar contention range most of the time. So, they will have the same likelihood of channel access and thus IEEE 802.11 DCF MAC protocol achieves time-based fairness among nodes with similar bit rates [9]. If all nodes use the same frame size, 802.11 MAC protocol also provides throughput-based fairness (i.e. equal throughput shares). However, throughput fairness maybe caused bandwidth underutilization in multi-rate wireless networks. Also there are two other types of fairness: max-min fairness and proportional fairness. It was proved that if all nodes have the same weight, then the proportional fairness will be equivalent to the time-based fairness in the networks with multi data rates. Moreover, it was proved that the max-min fairness and throughput-based fairness are equivalent under the same condition. It is also argued that the proportional fairness achieves trade-off between throughput and fairness in the network [20].

The standard 802.11 MAC protocol achieves max-min fairness in the utilization of the bandwidth. In multi data rate networks, the nodes with low data rate will consume more air-time and this differs from max-min fairness in the use of the bandwidth. Proportional fairness with respect to the use of the bandwidth is nearly equal to the max-min fairness with respect to the use of air-time [20].

Proportional fair scheduling at each access point (AP) was proposed in order to get the balance between the aggregate throughput and the fairness in serving all nodes [21]. Proportional fair scheduling also achieves the trade-off between fairness and efficiency if there is a single AP that supports multi-rates.

Proportional fairness at the access point is for the distribution of the bandwidth on nodes in proportion to their data rates, or assigning the time equally among them if there is a single AP and all nodes with similar priority. In the case of multiple AP, proportional fairness will work by maximizing the sum of logarithms of bandwidth allocated for each node [21].

Fair allocation was proposed [22] for coverage overlapping cells, in order to achieve the proportional fairness. Each new arrival mobile station (MS) will be linked to only one base station (BS). Under this constraint, Bin *et al.* [22] try to maximize the logarithm of total data rate for all mobile stations. Hence, every single BS must work on maximizing the logarithm of total data rate over its assigned MSs by allocating radio resources in a proportionally fair manner. They formulate general model of proportional fairness for overlapping coverage in multiple wireless cells.

The fair rate allocation problem was considered [23] to monitor the entire coverage area and maximize the total proportionally fairness in WSNs with regular topologies where the aforementioned problem was studied for WSNs with regular topologies that used slotted Aloha MAC layer. Narayanan *et al.* [23] conclude that, the best topology for the proportionally fair throughput is triangular, square and then hexagonal with respect to the growth size of the network, going from small, medium, to large respectively.

Proportional Fairness Backoff (PFB) scheme was proposed to overcome the funneling effects of the converge-cast patterns in WSNs [24] [25]. Sensors that are in the area around the sink will have larger number of packets than far away sensors. Because of this disproportionate number of accumulated packets, it is very important to decrease the collisions number as well as increasing the throughput to mitigate this funneling effect in the Intensity Region, the region of the funnel.

Yuanfang *et al.* [24] [25] proposed a new scheme, PFB which provides additional opportunities of channel access to the nodes that are closer to the sink in order to address the funneling effect problem.

This problem has many effects in WSNs such as reducing the amount of the data gathered by the sink, shortening the lifetime of the sensors, breaking the stability as well as decreasing the whole throughput of WSN.

The probabilistic approach [26] provides the proportional fairness without having to solve an optimization problem. Load estimation strategy is used in this approach to estimate the total traffic load for each node, and thus adjusting the contention window according to the difference between the current share of the channel and the required one. So, each node will get a share of the channel that is appropriate to its traffic load. Hence, proportional fairness will be achieved among nodes.

In the next Section, we provide the details of our proposed protocol: Fair MAC protocol (FMAC).

5. Methodology

Wireless sensor networks are usually used in the monitoring applications. When a certain event happens in the monitored area, a large number of sensors will need to transmit their data simultaneously. But because of the WSNs nature, it is possible to find heterogeneous nodes with different data rates. Nodes with lower data rate will occupy channel for longer duration; this will result into higher delays and degrades the whole throughput of the network.

Another issue is that because of this many-to-one data transfer model, all nodes send to the sink, it is important to increase the throughput and reduce the number of collisions in order to mitigate funneling effects in the intensity region, the region of the funnel around the sink. We intend to find a new scheme which will reduce the collisions and utilize the idle slots of time, as well as providing more opportunities to the nodes that have higher data rates. Ultimately, we should achieve the proportional fairness among all nodes in the network.

Fair MAC Protocol (FMAC)

We propose the Fair MAC protocol (FMAC) which is based on the idea of using the medium delay periods within the backoff relative to the data rates of the nodes. This will give more opportunities to access the channel for the fast nodes and thus addresses the performance anomaly problem. The use of the medium periods within the backoff reduces the collisions and utilizes the idle time slots.

In FMAC, when the nodes want to calculate the backoff exponential value (BE), they also retrieve the data rate in order to calculate the medium delay (MD) value. Based on the data rate value, it will choose the value of the medium period (MP). MP is the value that determines how the percentage of MD should be from the actual backoff delay (B_Time). After computing the value of MP, which will be between 10 and 25 for fast nodes and between 26 and 40 for low nodes, this value will be used directly to get the percentage of MD from the current backoff delay value. MD value ranges between 10% and 40% of the current backoff delay. Calculations are done as follows:

$$MP = 10 \rightarrow 40. \tag{2}$$

$$B\_Time = (2BE - 1) * UBP. \tag{3}$$

$$MD = (B\_Time)(MP)/100. \tag{4}$$

where B_Time is the actual backoff delay selected randomly by the node using the values: BE = 1, 2,3,4,5 and the unit backoff period (UBP).

The main stages of the proposed protocol are shown below in **Figure 2** where we only concentrate on the case when the node needs to apply the modified backoff algorithm.

Firstly, the node calculates the actual backoff delay (B_Time). Secondly, it retrieves its data rate internally. Considering the data rate to be high or low can vary depending on the type of application used which highlights the important of selecting the right threshold.

If the node has a high data rate, then MP value will be selected randomly from 10 to 25. Else, if the node has a slow data rate, then MP value will be chosen randomly between 26 and 40. Next, the value of MP will be used to get the value of the percentage MD. After the period MD has elapsed, the channel will be assessed. If the channel is found idle, then the node will start the sending process. While if the channel is not idle, the node will

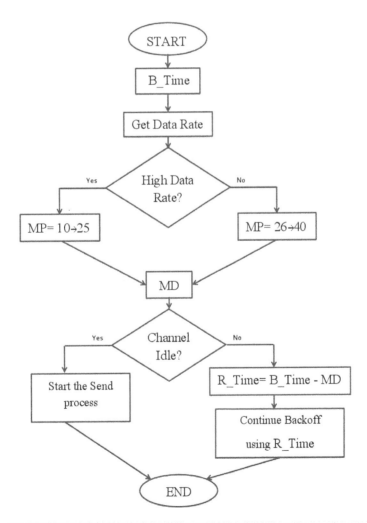

Figure 2. Flow chart of the main stages of FMAC.

continue the backoff process using the remaining time (R_Time), after subtracting the value of the elapsed MD, from the actual backoff delay (B_Time).

Assigning different values of MP for the nodes that have different data rates gives different opportunities for these nodes. In other words, the nodes that have a high data rate will be able to assess the channel after a relatively small delay (MD) from the actual backoff delay (B_Time). This means that if the collision occurs, the fast nodes will get more opportunity to access the channel after starting to apply the backoff process.

On the other hand, slow nodes will have a bigger value of MP and thus they will wait more time (MD) in order to be able to assess the channel. Of course, this will reduce their chances to access the channel, because it may be captured by the fast nodes.

FMAC is expected to contribute significantly in reducing the collisions because of the low probability of choosing the same random number for the backoff delay by any two nodes, and also choosing the same random number for the medium delay by these two nodes.

A very important point that must be taken into account here is that although FMAC gives more opportunity to access the channel for the fast nodes, it does not cause starvation for the slow nodes. This is because all these medium delay periods are actually applied within the main frame of the standard backoff process. FMAC enables both of the fast and slow nodes to assess the channel before the end of the actual backoff delay in order to utilize the idle slots of time and reduce the delay.

Another important point is that medium delay periods used in FMAC is similar to those periods defined in Improved Binary Exponential Backoff (IBEB) [27], but not the same. Since IBEB does not focus mainly on achieving proportional fairness. Also, the computation of interim periods in IBEB is done more than once and without considering the data rate of the nodes.

6. Simulation and Performance Analysis

We implement Fair MAC protocol (FMAC) discussed in Section 4 using Network Simulator NS-2 version 2.35 under Linux (Fedora 19).

Simulated area is 350 m × 350 m. It contains 11 static nodes deployed randomly without assuming densely deployment but with ensuring the connectivity of the network. One of the nodes is the sink node which is every time located at the middle of the area and it assumed that it has stronger capabilities than other nodes. The remaining nodes are sensor nodes which are sending their data only to the sink in a one hop manner.

We assume the applications that involve sink-oriented transmissions [28]. We consider the uplink direction of the transmissions in a saturated network; each node always has data to be sent.

Nowadays, there are many applications need higher data rates than those that were used in old WSNs. these new applications include: media streaming (audio and video data), target tracking, exploration of disaster and critical controlling [29]-[32]. So, we assume high data rates which are similar to those that provided by 802.11 protocol.

Figure 3 shows an example of the assumed simulation area. In this Figure, the random topology with 20 sensor nodes is shown. There are 14 fast nodes and 6 slow nodes. The number of slow and fast nodes was selected randomly.

FMAC protocol is employed as the protocol for MAC layer. The performance of FMAC protocol is compared every time with that of the standard 802.11 MAC protocol. The mechanism of sending of RTS/CTS is used to safely access the channel for both protocols.

We use different data rates to represent the fast nodes and the slow nodes selecting 2 Mbps and 1 Mbps for this purpose respectively; according to the rates which are provided by the standard MAC 802.11, because of comparison issues.

Constant Bit Rate (CBR), based on the UDP protocol, used as traffic source generator. The number of sent packets was varied by tuning the interval of CBR. However, all nodes have the same interval along the single experiment.

We increase the number of sensor nodes gradually from 10 to 50 nodes in order to test the scalability of our protocol. Although that the numbers of slow and fast nodes are chosen randomly, we ensure that at least 30% of all nodes are slow nodes. We choose this percentage in order to see the impact of slow nodes clearly. Also the number of sent packets was increased gradually from 200 to 1000 packets to ensure the performance of our protocol under different traffic loads.

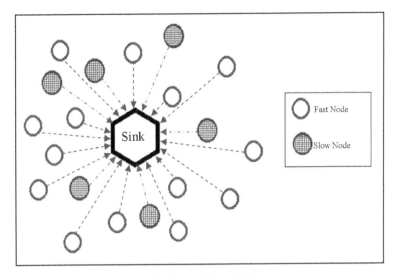

Figure 3. The virtual simulation area.

The time of simulation is 200 seconds and each experiment was done 30 times, then the average was calculated to produce the graphs for the results provided later. **Table 1** shows a summary of the important parameters of the simulation. The values of other parameters unchanged as defined in the files of NS-2.

The major matrices used in this research to evaluate the performance of our proposed protocol are as follows: number of collisions, energy consumption, end-to-end delay, throughput aggregation for the network as a whole and Jain's fairness index. We also make a comparison of the throughput for all the fast nodes with the throughput for all the slow nodes. The simulation results are provided in the next Section with Illustrative graphs.

7. Results and Discussion of Implications

All networks that are based on contention between nodes mainly experience the problem of collisions. As long as the number of collisions is increasing, the retransmission number is also increased, and thus more energy is consumed. Of course, this will affect the overall performance of the network, especially because of the limited resources in WSN.

On the other hand, as mentioned before, the performance anomaly, which is caused by slow nodes, will degrade the overall throughput of the network. So, we focus in this research on studying and analyzing these main related metrics.

7.1. Collisions

Although using of RTS/CTS mechanism solves the collision problem that is caused by the hidden terminal problem to some extent, there is still another reason for the collision. This reason is the probability of choosing the same number of the backoff interval time by two nodes that are in the range of each other. **Figure 4** shows the comparison of collisions number for FMAC protocol and the standard 802.11 protocol under different sizes of the network.

The average number of collisions was reduced clearly when FMAC protocol is used instead of the standard MAC 802.11 protocol with different number of sent packets. **Figure 4** shows that when the number of sent packets is increasing, the difference between FMAC and standard MAC 802.11 becomes bigger.

Collisions also tested under different loads; different numbers of sent packets, the results are shown in **Figure 5**. Although both protocols have a similar number of collisions with small number of nodes as in **Figure 5**, when increasing the number of nodes there is a noticeable difference for FMAC favor. There is about 4.33% reduction of the collisions number with FMAC protocol.

7.2. End to End Delay

End to end delay also has an effect on the performance of the network as a whole. **Figure 6** and **Figure 7** show

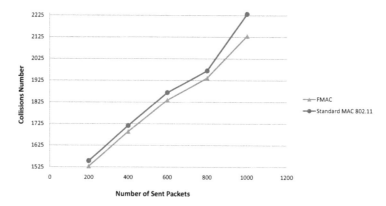

Figure 4. Comparison of collisions number for FMAC and standard MAC 802.11 (30 nodes).

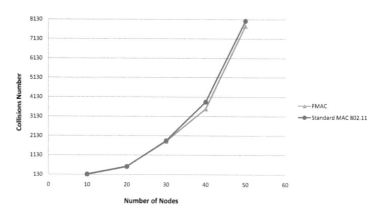

Figure 5. Comparison of collisions number for FMAC and standard MAC 802.11 (600 packets).

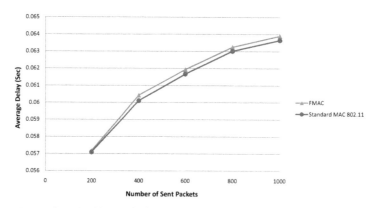

Figure 6. Comparison of end-to-end delay for FMAC and standard MAC 802.11 (40 nodes).

Table 1. Important parameters of the simulation.

Parameters	Values
MAC Layer Protocols	FMAC, Standard IEEE 802.11 MAC protocol
Routing Protocol	DSDV
Number of Sensor Nodes	10, 20, 30, 40, 50
Number of Packets	200, 400, 600, 800, 1000

the comparison of end-to-end delay for both FMAC protocol and the standard MAC 802.11 protocol under different loads.

We can see from **Figure 6** that end-to-end delay with FMAC protocol is slightly bigger than that with the standard MAC 802.11 protocol. However, the delay for both protocols is increased with the increase of the number of sent packets.

In **Figure 7**, we increase the number of nodes in the network and compare end-to-end delay for both two mentioned protocols.

We can see clearly that the delay with FMAC protocol is very similar to that with the standard MAC 802.11 protocol in all cases. However, the delay becomes a slightly bigger with the largest network size. FMAC increases the average end-to-end delay with only around 0.27%. As a result, such delay is insignificant in WSNs.

7.3. Energy Consumption

The secret of success for any protocol proposed to be applied in WSNs is the power saving.

In **Figure 8** we provide the comparison of the average consumed power per node for FMAC and the standard MAC 802.11 protocol under different number of sent packets. When increasing number of sent packets, nodes in FMAC protocol consume a slightly less power than the standard MAC 802.11 protocol. However, both protocols consume more energy with heavy load.

Figure 8 shows the increase in energy consumption becomes slower with high loads, while **Figure 9** shows the average consumed power per node for our proposed protocol and the standard MAC 802.11 protocol under different sizes of the network.

In **Figure 9** we see that the consumed energy by both protocols is also increased as long as the size of the

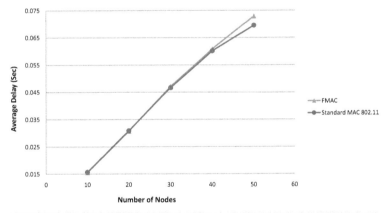

Figure 7. Comparison of end-to-end delay for FMAC and standard MAC 802.11 (200 packets).

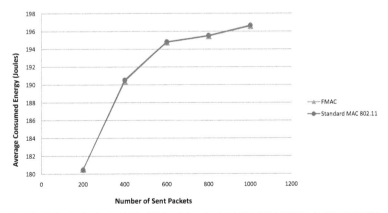

Figure 8. Comparison of energy consumption for FMAC and standard MAC 802.11 (20 nodes).

network increased. But, with significantly difference in favor of FMAC protocol. FMAC consumed less power than the standard MAC 802.11 protocol.

FMAC utilizes the idle time and reduces the number of retransmissions since the collisions were reduced. The average energy consumption is reduced by about 0.011% when applying FMAC. Although this is a small value, it considered sufficient to ensure that our protocol does not cause more energy consumption.

7.4. Aggregation Throughput

One of the main objectives of the proposed protocol is enhancing the aggregate throughput of the network as a whole. **Figure 10** presents the comparison of aggregate throughput for FMAC and the standard MAC 802.11 protocols with different number of sent packets. As **Figure 10** indicates when increasing the number of packets sent, FMAC protocol achieves greater throughput than the standard MAC 802.11 protocol. This means that the aggregate throughput of the network becomes larger when using FMAC protocol in a network of high load.

Aggregate throughput for FMAC and the standard MAC 802.11 protocols with different sizes of the network is shown in **Figure 11**. It shows that FMAC protocol still gives better throughput than the standard MAC 802.11 protocol. Of course, this significant difference in favor of FMAC is due to its ability to decrease the collisions.

Generally, FMAC protocol achieved an increase in the average aggregate throughput in all cases of about 1.9% higher than the standard MAC 802.11 protocol.

7.5. Proportional Fairness

WSN is usually used for monitoring purposes, which require gathering information from a large number of nodes in a fair manner as possible, while caring about throughput and aggregation throughput per node.

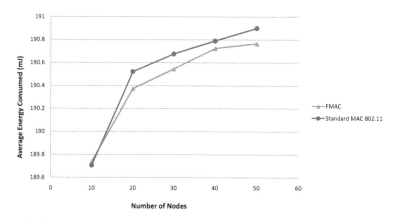

Figure 9. Comparison of energy consumption for FMAC and standard MAC 802.11 (1000 packets).

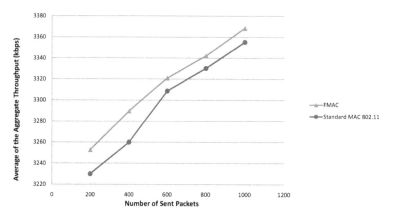

Figure 10. Comparison of aggregate throughput for FMAC and standard MAC 802.11 (50 nodes).

We are focusing on achieving proportional fairness between slow nodes and fast nodes. So, we calculate the average throughput for all fast nodes and then we compare this average with the average throughput for all slow nodes. Every time there is about 30% of nodes that have slow data rate.

Figure 12 shows this comparison for the standard 802.11 MAC protocol with different number of nodes, while **Figure 13** shows the same thing but for FMAC protocol. The average of the throughput for all slow nodes is still larger than the average of the throughput for all fast nodes while the size of the network is increasing with the standard MAC 802.11 protocol as shown in **Figure 12**. Despite this, the slow nodes constitute only about 30%

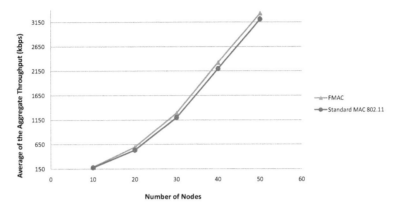

Figure 11. Comparison of aggregate throughput for FMAC and standard MAC 802.11 (400 packets).

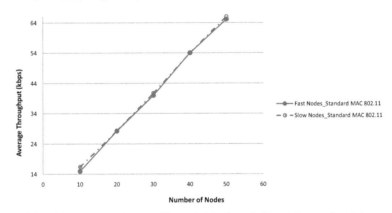

Figure 12. Comparison throughput of fast and slow nodes in standard MAC 802.11 (200 packets).

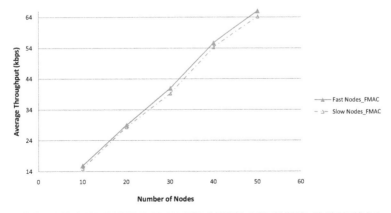

Figure 13. Comparison throughput of fast and slow nodes in FMAC (200 packets).

of the total number of nodes in the network. On the other hand, we can see in **Figure 13** that the average of throughput for all slow nodes clearly becomes less than the average of the throughput for all fast nodes when applying FMAC protocol while the size of the network is increasing.

The same comparison between these two protocols is shown in **Figure 14** and **Figure 15** respectively, but this time with a different number of sent packets. In **Figure 14**, we can see that the difference in the average throughput for all slow nodes and the average throughput for all fast nodes becomes more significance when the number of sent packets is increased with the standard MAC 802.11 protocol. Again, the slow nodes that comprise about only 30% of the total number of nodes in the network have a higher average of throughput.

We can see exactly the opposite in **Figure 15**; the average of the throughput for all slow nodes significantly becomes less than the average of the throughput for all fast nodes when the number of sent packets is increased with FMAC 802.11 protocol. FMAC protocol achieves proportional fairness, between slow and fast nodes, more than the standard MAC 802.11 in all cases. But the difference is noticed more clearly with increasing of number of sent packets. This is due the fact that FMAC protocol gives more chance for the fast nodes to access the channel.

7.6. Fairness Index

Fairness index measures how the channel is shared equally by all nodes. It is concerned with the minimum number of transmitted packets by any individual node in the network relative to the maximum number of transmitted packets by any of the nodes in the network [33]. Jain's index gives the fairness criterion taking into account all the nodes in the network [34]. Hence, the higher value of this index, the better fairness between all nodes. Jain's fairness index is given by [35]:

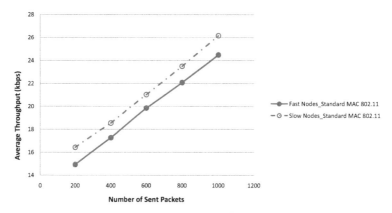

Figure 14. Comparison throughput of fast and slow nodes in standard MAC 802.11 (10 nodes).

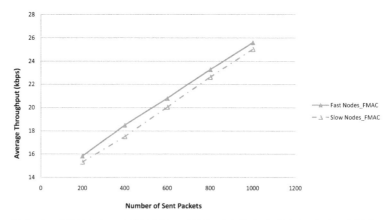

Figure 15. Comparison throughput of fast and slow nodes in FMAC (10 nodes).

$$f(x) = \frac{\left(\sum_{i=1}^{n} x_i\right)^2}{\sum_{i=1}^{n} x_i^2}, \quad x_i \geq 0. \tag{5}$$

where n is the number of all contending nodes and x_i is share of the allocation which given for the ith node.

We compute Jain's fairness index twice. The first time is for FMAC protocol with different sizes of the network and for the average number of sent packets. **Figure 16** shows the comparison of this index for FMAC and the standard MAC 802.11 protocol.

The second time is shown in **Figure 17**, which is also for FMAC protocol but with different loads in the network and for the average number of nodes with the comparison of the standard MAC 802.11 protocol.

From both **Figure 16** and **Figure 17**, it is clear that the value of Jain's fairness index for FMAC protocol is higher than the standard MAC 802.11 protocol in all cases. As the load in the network increases, the difference between the values of the index for both two protocols becomes bigger. Also we can see in **Figure 16** that with the largest network size, the value of the index for FMAC protocol becomes very close to 1, which is the maximum value of Jain's fairness index and it indicates a higher achieved fairness.

8. Conclusion and Future Work

The main objective of the proposed algorithm is to design a new MAC protocol that achieves proportional fairness between nodes with different data rates in WSNs. Fair MAC protocol (FMAC) aims to enhancing the aggregate throughput as well as achieving the proportional fairness between all nodes in the network by reducing the collisions number and utilizing the idle slots of time.

The core idea of FMAC protocol is that each node computes a medium delay period in proportion to its data rate, directly after the computation of the actual backoff delay period. Then, when the medium period has

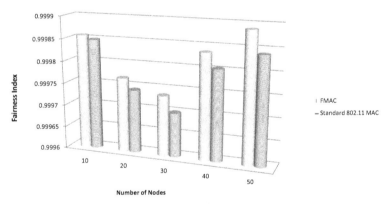

Figure 16. Comparison of Jain's fairness index of FMAC and standard MAC 802.11 (average load).

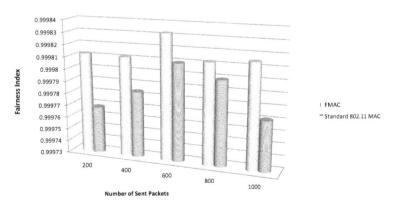

Figure 17. Comparison of Jain's fairness index of FMAC and standard MAC 802.11 (average number of nodes).

elapsed, it will check the channel to be idle. If the channel is busy, the node will continue the backoff process with remaining delay of the actual period. This will achieve proportional fairness and maximize the aggregate throughput by giving more probability to access the channel for the fast nodes. FMAC protocol is compatible with the standard IEEE 802.11, only small changes are required. However, the proposed protocol is also applicable in WLANs because it does not mainly depend on the nature of WSNs.

The experimental results show that when using FMAC the transmission becomes faster with less power consumption due to the better utilization of idle time slots. Also the average number of collisions is reduced by 4% and the average aggregation throughput increased by 1.9%. Using Jain's fairness index, FMAC protocol obtains higher values. Thus, experimental results indicate that FMAC protocol achieves its main objectives.

The experiments also reveal some interesting future work such as achieving more aggregated throughput by adjusting the interim delay periods more specifically. Also we may compare our protocol against other protocols in existence today, analytically and/or using the simulation. Another possible future direction would be to perform the proposed protocol in different environments or using real world experiments.

Acknowledgements

This paper was supported by NSTIP strategic technologies program project number 10-INF1184-02 in the Kingdom of Saudi Arabia.

References

[1]   Xu, N. (2002) A Survey of Sensor Network Applications. *IEEE Communications Magazine*, **40**, 102-114.

[2]   Corke, P., Wark, T., Jurdak, R., Wen, H., Valencia, P. and Moore, D. (2010) Environmental Wireless Sensor Networks. *Proceedings of the IEEE*, **98**, 1903-1917. http://dx.doi.org/10.1109/JPROC.2010.2068530

[3]   Wang, Z.Q., Yu, F.Q., Tao, L.Q. and Zhang, Z.S. (2011) A Fairness Spatial TDMA Scheduling Algorithm for Wireless Sensor Network. 12*th International Conference on Parallel and Distributed Computing, Applications and Technologies* (*PDCAT*), Gwangju, 20-22 October 2011, 348-353.

[4]   Sridharan, A. and Krishnamachari, B. (2007) Maximizing Network Utilization with Max-Min Fairness in Wireless Sensor Networks. 5*th International Symposium on Modeling and Optimization in Mobile, Ad Hoc and Wireless Networks and Workshops*, Limassol, 16-20 April 2007, 1-9. http://dx.doi.org/10.1109/WIOPT.2007.4480030

[5]   Pei, H., Li, X., Soltani, S., Mutka, M.W. and Ning, X. (2013) The Evolution of MAC Protocols in Wireless Sensor Networks: A Survey. *IEEE Communications Surveys & Tutorials*, **15**, 101-120. http://dx.doi.org/10.1109/SURV.2012.040412.00105

[6]   Kumar, B., Yadav, R.K. and Challa, R.K. (2010) Comprehensive Performance Analysis of MAC Protocols for Wireless Sensor Networks. *International Conference on Computer and Communication Technology* (*ICCCT*), Allahabad, 17-19 September 2010, 342-347. http://dx.doi.org/10.1109/ICCCT.2010.5640504

[7]   Patil, U.A., Modi, S.V. and Suma, B. (2013) Analysis and Implementation of IEEE 802.11 MAC Protocol for Wireless Sensor Networks. *International Journal of Engineering Science and Innovative Technology* (*IJESIT*), **2**, 278-284.

[8]   Singh, U.K., Phuleriya, K.C. and Laddhani, L. (2012) Study and Analysis of MAC Protocols Design Approach for Wireless Sensor Networks. *International Journal of Advanced Research in Computer Science and Software Engineering*, **2**, 79-83.

[9]   Duda, A. (2008) Understanding the Performance of 802.11 Networks. 19*th International Symposium on Personal, Indoor and Mobile Radio Communications*, Cannes, 15-18 September 2008, 1-6. http://dx.doi.org/10.1109/PIMRC.2008.4699942

[10] Kabara, J. and Calle, M. (2012) MAC Protocols Used by Wireless Sensor Networks and a General Method of Performance Evaluation. *International Journal of Distributed Sensor Networks*, **2012**, Article ID: 834784. http://dx.doi.org/10.1155/2012/834784

[11] Jain, S. and Mahajan, R. (2000) Wireless LAN MAC Protocols.

[12] Nadeem, T. and Ashok, A. (2005) Performance of IEEE 802.11 Based Wireless Sensor Networks in Noisy Environments. 24*th IEEE International Conference on Performance, Computing, and Communications*, Phoenix, 7-9 April 2005, 471-476. http://dx.doi.org/10.1109/PCCC.2005.1460615

[13] Zhai, H.Q., Kwon, Y. and Fang, Y.G. (2004) Performance Analysis of IEEE 802.11 MAC Protocols in Wireless LANs. *Wireless Communications and Mobile Computing*, **4**, 917-931. http://dx.doi.org/10.1002/wcm.263

[14] Bononi, L., Conti, M. and Gregori, E. (2000) Design and Performance Evaluation of an Asymptotically Optimal Backoff Algorithm for IEEE 802.11 Wireless LANs. *Proceedings of the 33rd Annual Hawaii International Conference on*

System Sciences, Maui, 4-7 January 2000, 10. http://dx.doi.org/10.1109/HICSS.2000.926987

[15] Khalaj, A., Yazdani, N. and Rahgozar, M. (2007) Effect of the Contention Window Size on Performance and Fairness of the IEEE 802.11 Standard. *Wireless Personal Communications*, **43**, 1267-1278. http://dx.doi.org/10.1007/s11277-007-9300-5

[16] Weinmiller, J., Woesner, H. and Wolisz, A. (1996) Analyzing and Improving the IEEE 802.11-MAC Protocol for Wireless LANs. *Proceedings of the 4th International Workshop on Modeling, Analysis, and Simulation of Computer and Telecommunication Systems, MASCOTS'96*, San Jose, 1-3 February 1996, 200-206.

[17] Chen, S. and Zhang, Z. (2006) Localized Algorithm for Aggregate Fairness in Wireless Sensor Networks. *Proceedings of the 12th Annual International Conference on Mobile Computing and Networking*, Los Angeles, 24-29 September 2006, 274-285. http://dx.doi.org/10.1145/1161089.1161121

[18] Wu, J.H. and Luo, J. (2012) Research on Multi-Rate in Wireless Sensor Network Based on Real Platform. *2nd International Conference on Consumer Electronics, Communications and Networks (CECNet)*, Yichang, 21-23 April 2012, 1240-1243.

[19] Yang, T., Kulla, E., Oda, T., Barolli, L., Younas, M. and Takizawa, M. (2012) Performance Evaluation of WSNs Considering MAC and Routing Protocols Using Goodput and Delay Metrics. *15th International Conference on Network-Based Information Systems (NBiS)*, Melbourne, 26-28 September 2012, 341-348.

[20] Jiang, L.B. and Liew, S.C. (2005) Proportional Fairness in Wireless LANs and Ad Hoc Networks. *IEEE Wireless Communications and Network Conference (WCNC)*, **3**, 1551-1556.

[21] Li, L., Pal, M. and Yang, Y.R. (2008) Proportional Fairness in Multi-Rate Wireless LANs. *IEEE INFOCOM 2008, The 27th Conference on Computer Communications*, Phoenix, 13-18 April 2008, 1004-1012.

[22] Chen, B.B. and Chan M.C. (2006) Proportional Fairness for Overlapping Cells in Wireless Networks. *IEEE 64th Vehicular Technology Conference*, Montreal, 25-28 September 2006, 1-5.

[23] Narayanan, S., Jun, J.H., Pandit, V. and Agrawal, D.P. (2011) Proportionally Fair Rate Allocation in Regular Wireless Sensor Networks. *IEEE Conference on Computer Communications Workshops*, Shanghai, 10-15 April 2011, 549-554. http://dx.doi.org/10.1109/INFCOMW.2011.5928874

[24] Chen, Y.F., Li, M.C., Wand, L., Yuan, Z.X., Sun, W.P., Zhu, C.S., Zhu, M. and Shu, L. (2009) A Proportional Fair Backoff Scheme for Wireless Sensor Networks. *IEEE 6th International Conference on Mobile Adhoc and Sensor Systems*, Macau, 12-15 October 2009, 971-976.

[25] Chen, Y.F., Li, M.C., Shu, L., Wang, L. and Hara, T. (2012) A Proportional Fairness Backoff Scheme for Funnelling Effect in Wireless Sensor Networks. *Transactions on Emerging Telecommunications Technologies*, **23**, 585-597. http://dx.doi.org/10.1002/ett.2516

[26] Chakraborty, S., Swain, P. and Nandi, S. (2013) Proportional Fairness in MAC Layer Channel Access of IEEE 802.11s EDCA Based Wireless Mesh Networks. *Ad Hoc Networks*, **11**, 570-584. http://dx.doi.org/10.1016/j.adhoc.2012.08.003

[27] Khan, B.M., Ali, F.H. and Stipidis, E. (2010) Improved Backoff Algorithm for IEEE 802.15.4 Wireless Sensor Networks. 2010 *IFIP, Wireless Days (WD)*, Venice, 20-22 October 2010, 1-5.

[28] Akan, O.B. and Akyildiz, I.F. (2005) Event-to-Sink Reliable Transport in Wireless Sensor Networks. *IEEE/ACM Transactions on Networking*, **13**, 1003-1016. http://dx.doi.org/10.1109/TNET.2005.857076

[29] Jagadeesha, R. and Alfandi, O. (2013) Implementation of WSN Protocol on a Heterogeneous Hardware. *Journal of Engineering Science and Technology*, **8**, 521-539.

[30] Zhu, N.H., Du, W., Navarro, D., Mieyeville, F. and Connor, I.O. (2011) High Data Rate Wireless Sensor Networks Research. *Proceedings of 14ème Journées Nationales du Réseau Doctoral de Micro et Nanoélectronique (JNRDM 2011)*, Paris, 23-25 May 2011.

[31] Zhu, N.H. and O'Connor, I. (2013) Performance Evaluations of Unslotted CSMA/CA Algorithm at High Data Rate WSNs Scenario. *9th International Wireless Communications and Mobile Computing Conference (IWCMC)*, Sardinia, 1-5 July 2013, 406-411.

[32] Kohvakka, M., Arpinen, T., Hännikäinen, M. and Hämäläinen, T.D. (2006) High-Performance Multi-Radio WSN Platform. *Proceedings of the 2nd International Workshop on Multi-Hop Ad Hoc Networks: From Theory to Reality*, Florence, 26 May 2006, 95-97.

[33] Nithya, B., Mala, C. and KumarB, V. (2012) Simulation and Performance Analysis of Various IEEE 802.11 Backoff Algorithms. *Procedia Technology*, **6**, 840-847. http://dx.doi.org/10.1016/j.protcy.2012.10.102

[34] Bin Sediq, A., Gohary, R.H. and Yanikomeroglu, H. (2012) Optimal Tradeoff between Efficiency and Jain's Fairness Index in Resource Allocation. *IEEE 23rd International Symposium on Personal Indoor and Mobile Radio Communications (PIMRC)*, Sydney, 9-12 September 2012, 577-583.

[35] Jain, R., Chiu, D.-M. and Hawe, W.R. (1984) A Quantitative Measure of Fairness and Discrimination for Resource Allocation in Shared Computer System: Eastern Research Laboratory. Digital Equipment Corporation.

7

Determining Requirements for an Optimal Ad-Hoc Multicast Protocol in Mobile Health Care Training Systems

Anis Zarrad, Ahmed Redha Mahlous

Department of Computer Science and Information Systems, Prince Sultan University, Riyadh, Saudi Arabia
Email: azarrad@psu.edu.sa, armahlous@psu.edu.sa

Abstract

Nowadays, Health Care Training-based System (HCTS) is a vital component in the education and training of health care in 3D Virtual Environment (VE). The practice of HCTS continues to grow at rapid pace throughout all of the healthcare disciplines, however research in this field is still in its early stage. Increasingly, decision makers and developers look forward to offer more sophisticated, much larger, and more complex HCTS to serve the desired outcome and improve the quality and safety of patient care. Due to the rapidly increasing usage of personal mobile devices and the need of executing HCTS applications in environments that have no previous network infrastructure available, Mobile Health Care Training-based System (MHCTS) is an expected future trend. In such systems, medical staff will share and collaborate in a 3D virtual environment through their mobile devices in an ad-hoc network (MANET) in order to accomplish specific missions' typically surgical emergency room. Users are organized into various groups (Radiologists, Maternity departments, and General surgery etc...), and need to be managed by a multicast scheme to save network bandwidth and offer immersive sense. MHCTS is sensitive to networking issues, since interactive 3D graphics requires additional load due to the use of mobile devices. Therefore, we need to emphasize on the importance and the improvement of multicast techniques for the effectiveness of MHCTS and the management of collaborative group interaction. Research so far has devoted little attention to the network communication protocols design of such systems which is crucial to preserve the sense of immersion for participating users. In this paper, we investigate the effect of multicast routing protocol in advancing the field of Health care Training-based System to the benefit of patient's safety, and health care professional. Also, we address the issue of selecting a multicast protocol to provide the best performance for a particular e-health system at any time. Previous work has demonstrated that multicast operates at least as efficiently as traditional MAODV. A comprehensive analysis about various ad-hoc multicast routing protocols is proposed. The selection key factors for the right protocol for MHCTS applications were safety and robustness. To the best of our knowledge, this work will be the first initiative involving systematic literature reviews to identify a research gate for the use of multicast protocol in health care simulation learning community.

Keywords

Ad-Hoc, Multicast Protocol, Health Care Training System, 3D Virtual Environment, E-Health

1. Introduction

Mobile Health Care Training-based System (MHCTS) using virtual reality environment and telecommunication technologies is becoming important to both economic and patient's life safety. MHCS could be defined as a system that allows multiple trainees distributed geographically, to practice proper medical decision in the virtual environment, using their mobile devices in order to interact and collaborate with virtual patients. Such system let the trainee touch, feel, and manipulate virtual patients in safe way. Each trainee is represented by a graphical embodiment called avatar [1]. For the purpose of health care professional, the possibility to access detailed 3D virtual environment (VE) provides a promising alternative to traditional training. The unguided wireless medium and surrounding physical environment significantly disturb the radio transmissions, resulting in relatively unreliable communication channels. As a result, providing mobile healthcare VE simulator where conventional infrastructure-based communication facilities are not available, poses great challenges as more attention to network communication performance and 3D data exchange. Signal strength can easy degrades with distance between nodes. Therefore, there is a need to control network performance and consider the limited storage capabilities. Also, a specific problem related to the storage and performance capacity may occur, since users use their mobile devices with different characteristics to participate in MHCTS.

Currently, most of the health care simulators [2]-[4] are developed with desktop idea in mind and wired networks, thus hardware capabilities and network communication are not an issue. Health domain is largely benefit from collaborative virtual environment and wired network to provide the possibility of practicing medical procedures and share experiences with a remote tutor and other students [5]. In mobile health care virtual environment it is important to feed participants' devices only with necessary 3D data to display the user's viewpoint based on specific time and location in the VE. For example only avatars in the surgery room need to view the details about the room content. Avatars walking in the hospital corridor should not see what is inside the room. Nurses in the reception desks interact only with avatars in their area and can access their online data. Also, privacy is a central concern when dealing with capturing technologies for patients and can easily discomfort them.

To the researchers' knowledge, no attempt, except for work by Boukerch *et al.* [6], has been made to apply multicast protocol in e-health. Several steps have been made toward 3D rendering, streaming and 3D graphical display. However, network communication performance when dealing with VE over Ad-hoc networks has never been considered as an important concern. Emergency response efforts in critical situation are a must. For example in a car crash accident, wounded people are taken to the local hospital where they are forwarded directly to ER room. **Figure 1** shows an ER scenario in a hospital reception area.

An initial assessment and management of critically injured people require a rapid intervention with the following actions:

• Preparation of a resuscitation area and necessary monitoring equipments.
• Immediate call of medical staff.
• Prompt laboratory and radiology backup.

Once patients are received, immediately medical staff can start simulations to control and manage the ER wounded situation. All medical staff (nurses, doctors, Radiologists etc...) need to join the VE and collaborate together to deal with this situation. Surgeon need to manipulate virtual tissues and 3D anatomical models of organs through the same surgical tool usually used in real situation, while viewing images interactions on his mobile devices. 3D data and other shared data should be exchanged between participants in the VE.

MHCTS is a network delay- and bandwidth-sensitive application, especially in terms of message updates that reflect users' actions in the Virtual Environment (VE). Any interaction with the shared environment should be received in an acceptable delay to maintain the feeling of immersion and guarantee reality. In typical applications,

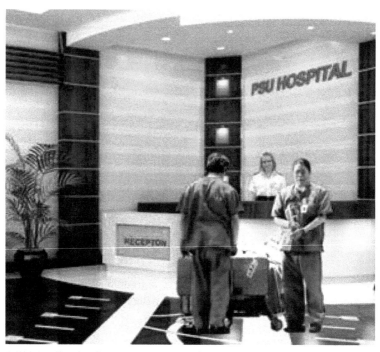

Figure 1. Emergency scenario.

network latency tolerance [7] usually varies between 10 ms to 1000 ms. However, it is argued in [8] that network latency in VE applications should not exceed 100 ms to preserve the immersion sense, but may reach a maximum of 200 ms [9] in some cases.

Multicast can be a primitive solution to cope with network delay and messages filtering. Only interested people will receive specific data. There is no need for nurses Avatars to receive data about the surgical procedure; only surgeons Avatars will receive the required data. Designing a multicast protocol to handle the network requirements and the sense of immersive for Mobile Health care Training-based (MHCTS) systems is a challenging task and requires expertise in systems engineering, robotics engineering, computer science, and medicine. In our previous work [6] we developed a multicast protocol based on Gnutella peer-to-peer networks to support mobile virtual environment. In this paper we work on providing a literature reviews for potential Multicast protocols in ad-hoc networks that can be applied in Mobile medical health care simulation. Our measurements and analysis of multicast protocols are driven by three primary concerns. The first one is the sensitivity and satisfaction of robustness because MHCTS should work under difficult circumstances (avoiding single points of failure). Robustness is an indicator for the resilience to errors and misinformation such as error prevention, fault detection, fault tolerance, redundancy, recovery and restart blocks. The second one is handling mobility and keeping nodes connected. The last concern has to deal with the decision of whether to propose a new protocol to meet MHCTS objectives or the use of an existing one.

The remainder of the paper is organized into three main sections. Section 2 provides a general idea about some used Multicast routing protocols over Ah-hoc network. Section 3 presents existing Multicast protocols applied in medicine. Section 4 concludes and presents future direction for the development of the new protocol.

2. Related Work a Short Survey

In recent years, a number of multicast protocols for ad hoc networks have been proposed. Most multicast communications in the internet involves the use of routing trees. The fundamental approach consists of creating a routing tree for a group of nodes (routers) in order to communicate efficiently. Thus, packets or message sent to all the routers in the tree traverses each router and link in the tree only once. Creating and using multicast routing trees have been extensively studied in Mobile ad hoc network and Wireless Sensor network (WSN). In order to meet the MHVTS real time requirement we classify proposed multicast protocols based on their robustness satisfaction, and multicast efficiency in order .

Multicast Operation of the Ad-hoc On-Demand Distance Vector Routing Protocol [10] MAODV is perhaps the most well know multicast protocol that represent a shared-tree-based protocol that is extended from AODV [11]. In MAODV, all members of a multicast group are formed into a tree, and the root of the tree is the group leader (first node join the tree). Multicast data packets are propagated among the tree. With the unicast route information of AODV, MAODV constructs the shared tree more efficiently and has low control overhead. Due to node mobility a tree partition may happen. A group member Q whose group leader has a lower IP address than any other group leader will inform its group leader to stop the leader's role. Node Q then sends a message to ask the group leader with the highest IP address to be the new group leader of the final merged tree. Advantages: With the unicast route information, the multicast tree can be constructed more quickly and efficiently. Disadvantages: The group leader continues flooding Group Hello messages even if no sender for the group exists.

Authors in [12] summarize and comparatively analyze the routing mechanisms of various existing multicast routing protocols according to the characteristics of mobile Ad Hoc network. From those protocols we find Ad Hoc Multicast Routing Protocol Utilizing Increasing Id-numbers (AMRIS) [13]. It is an on-demand shared-tree-based protocol which dynamically assigns every node in a multicast session an id-number known as msm-id. The msm-id provides each node with an indication of its "logical height" in the multicast delivery tree. Excluding the root node, each other node should have one parent that has a logical height (msm-id) that is smaller than it. A special node that is known as Sid with the smallest msm-id initiates initialization phase (multicast session) by broadcasting a NEW-SESSION message, in which each participant receiving the NEW-SESSION calculates its initial msm-id dynamically based on the value found in the message. If nodes receive multiple NEW-SESSION messages (for the same multicast session), they will only chose one message to process. Thus, the NEW-SESSION message thus travels in an expanding ring fashion outwards from Sid. In the initial phase, nodes closer to Sid will generally have smaller msm-ids than those further away.

These id-numbers help nodes know which neighbors are closer to the Sid and this reduces the cost to repair link failures (e.g. due to mobility terrain) and re-join the delivery tree in a localized fashion without causing permanent routing loops. One of the key features of AMRIS is on the way it provides routing information to other nodes. It doesn't use a unicast routing protocol, AMRIS maintains a Neighbor-Status table which stores the list of existing neighbors and their msm-ids.

The msm-ids allow nodes that have broken off from the delivery tree (e.g. due to mobility, terrain) to re-join the delivery tree in a localized fashion without causing permanent routing loops. One of the key features of AMRIS is that it does not depend on the unicast routing protocol to provide routing information to other nodes. AMRIS maintains a Neighbor-Status table which stores the list of existing neighbors and their msm-ids. Each node sends a periodic beacon to signal their presence to neighboring nodes. The beacon contains the msm-ids that each node presently has. However some disadvantages are present in AMRIS such as joining and re-joining of a node may take long time and waste much bandwidth since each node tries potential parent nodes arbitrarily and the usage of periodic beacons consumes bandwidth. Other protocol analyzed by [12], is the Lightweight Adaptive Multicast protocol (LAM) [14]. It provides services by using of the CBT (Core Based Tree) Algorithm and the TORA (Temporally-Ordered routing Algorithm). Similar with CBT, for each multicast group, LAM constructs a shared multicast tree whose center is the core node.

Two variables are maintained by nodes in LAM: POTENTIAL-PARENT and PARENT, and two lists, POTENTIALCHILD-LIST and CHILD-LIST. The PARENT variable is used to remember the parent node in the multicast tree, while the CHILD-LIST stores identities of one-hop child nodes in the multicasting tree. The POTENTIAL data structure will be used when the nodes are in a "join" or "rejoin" state. In mobile Ad Hoc network environment LAM has bad robustness. Another multicast WSN protocol can be found in literature. In [15] authors started by presenting a classification of the most common WSN multicast procedures depending on the way a target group is identified by the means of geographic position. They proposed a new protocol named Dijkstra-based Localized Energy-Efficient Multicast Algorithm (DLEMA) which is based on Localized Energy-Efficient Multicast Algorithm (LEMA) [16]. The first phase consist on building Minimum spanning Tree based (MST) on Kruskal's algorithm [17] seen from perspective of $s$source—a node currently routing a message. A member of a multicast group is added to MST only if at least one neighbor of s provides Euclidean advance towards that destination, or a perimeter relay node can be found.

For all destinations meeting the criteria, the current node becomes the root of the tree with edge costs reflecting the distance between nodes. Sensors constituting multicast group are the leaves or intermediate nodes. The only additional information about network topology available to s is the location of the nodes operating within its radio range. Therefore, the current node selects relay neighbors that are in the closest possible geographical proximity of each subset of destinations. In the second phase, LEMA uses Dijkstra's algorithm [18] to determine Energy Shortest Paths (ESPs) leading to neighbors that provide maximum geographical advance towards desired destinations. The procedure is based on the observation that ESP may consist of nodes not providing Euclidean advance but still minimizing total energy required to transport message from s to the end of the path. LEMA uses a localized source routing technique. A path determined in current node is stored in Source Routing Header (SRH) added to the message, while each sensor receiving a message with SRH should follow the requested route. Therefore, nodes that do not provide direct geographical advance can be designated as relays on the Energy Shortest Path to increase energy conservation of the network.

Unlike LEMA, DLEMA uses Dijkstra's algorithm to calculate shortest path tree (SPT) in its first phase. A performance evaluation of DLEMA, LEMA and Cone-based Forwarding Area Multicast tree (CoFAM) [19] was conducted through an extensive simulation. The results showed the outperformance of DLEMA over the other two protocols. In [20] authors present a receiver-initiated protocol called Core Assisted Mesh Protocol (CAMP). The multicast routing consists of building a shared mesh for each multicast group. It assumes that an underlying unicast routing protocol provides correct distances to known destinations. CAMP ensures that the shortest paths from receivers to sources (called reverse shortest paths) are part of a group's mesh. The number of packets coming from the reverse path for a sender indicates whether the node is on the shortest path. CAMP extends the basic receiver-initiated approach introduced in the core-based tree (CBT) protocol [21] for the creation of multicast trees to enable the creation of multicast meshes. CAMP advantages are: First, It constructs a mesh without control packet flooding. Second Node joining and shortest path procedure incurs very low overhead. However in the other hand CAMP has to rely on certain unicast routing protocols and offer high storage overhead.

3. Preliminaries: Multicast Routing Protocols Requirements for Mobile Health Care Training-Based System

In Mobile Health Care Training-based System users can interact and collaborate together. Users are assembled into groups based on their interest in the virtual environment. Users' immersive sensation is a critical issue that must be considered when designing such applications to offer certain realism. Therefore, MHCTS applications are considered to be sensitive to networking issues. During a higher transmission rates due to the use additional loads of 3D graphical data, it is hard to maintain real-time requirements and offer better performance. In general a multicast group is composed of senders and receivers. For connecting senders and receivers, each protocol constructs either a tree or a mesh as the routing structure. There are some nodes called forwarding nodes in the routing structure that are not interested in multicast packets but act as routers to forward them to receivers. Group members (senders and receivers) and forwarding nodes are also called tree or mesh nodes depending on the routing structure. In the routing structure, a node is an upstream (downstream) node of another node if it is closer to (farther away) the root of the tree. If the two nodes belong to the same link, the upstream (downstream) node is also called the parent (child) of the other node.

a) E-Health Requirements

E-Health systems have been well researched and studied for many numbers of years [22]. Recent advancements in mobile technologies have put a mobility requirement on traditional eHealth requirements. *Mobility* does offer more magnitude to the system, specifically the ability to work autonomously in a mobile environment without losing interest to the application. Classical eHealth requirements are: *Efficient communication mechanism* is crucial to reduce the network traffic generated by users' interaction messages, so that *large scalability* can be supported without affecting the overall performance of the proposed system without losing interest in the application. *Adaptability* should be also considered when developing multicast protocol, users may participate with different mobile devices having different criteria and capabilities such as, hardware, operating system, storage system, network capacity etc..., and therefore resulted protocol must be well adapted to different devices. In mobile environments, adaptability is considered a fundamental requirement due to the limitations of certain mobile devices.

In such system medical staffs collaborate toward specific objective. The state of the virtual environment and 3D data must be maintained during the whole mission and carried out even when user owner leave the mission. Persistency is important especially in eHealth system, for example when a nurse gives a medicine to a patient, the medical report should be updated. Thus, when the nurse finishes her shift and leaves the virtual environment, new comer is informed. Persistency can be achieved in multicast protocol by assigning backup nodes in the network to store all relevant actions and data for the mission. The last requirement we believe it important is the *Operability*, the system use standards and patterns in order to allow interactions between heterogeneous systems.

Attempting to build e-Health systems with all of the aforementioned requirements require immense efforts. In this work, we diminish the boundary requirements so as to consider only the mobility, and efficient network communication. The connectivity of the mobile nodes, route setup and repair time are the major factors that affect network performance.

b) Unicast versus Multicast versus Broadcast Routing

In the literature existing routing techniques can be classified either as Multicast, broadcast and unicast. Unicast is defined as a packet needs to go from a single source to a single destination. The communication is from a single node to another single node. There is one device transmitting a message destined for one receives. Contrary in multicast routing the packet needs to go from a single source to several destinations in a given address range not defined by any standard IP address and mask combination. This broadcast could reach all hosts on the subnet, all subnets, or all hosts on all subnets. Current routers, during the broadcast communication block IP broadcast traffic and restrict only to the local subnet. Multicast communication uses a distinct set of addresses. In broadcast routing the packet needs to go from a single source to every receiver exist in the network. The scope of the broadcast is limited to a broadcast domain. Cleary Multicast protocol can be an alternative solution to handle the e-Health application, because major critical activities are classified by department, and groups. Example surgery group, emergency departments etc...Also network bandwidth usability is more efficient because of multiple streams. Therefore, routing productivity is increased.

In [2] authors described the importance of network issue as a multicast protocol when using medical simulation in peer-to-peer network to achieve better realism during simulations and improve the immersive sense for all users participating in collaborative virtual environment. The proposed protocol is implemented in the collaborative module of the CyberMed VR framework. Greenhalgh *et al.* [3] describe a multicast protocol for Large Scale Collaborative Virtual Environments where the source node transmits only one packet information to a specific group of interested receivers. Instead of sending N separate but identical packets to each of N receivers, one multicast packet can be sent and multicast guarantees it will reach all N receivers. Therefore better scalability and decrease the use of the network bandwidth on CVEs, which allow better Quality and experience of the VE. Boukerch *et al.* [4] proposed an alternative solution of collaborative, haptic, audio and visual environments (C-HAVE) in order to cope with network delay, scalability, reliability and synchronization problem when the users are geographically distributed. A hybrid solution that incorporate four main protocols (the synchronous collaboration transport protocol (SCTP), the selective reliable transmission protocol (SRTP), the reliable multicast transport protocol (RMTP) and the scalable reliable multicast (SRM)) is proposed for Brain Tumor Tele-Surgery application.

4. Discussion

So far, the mobile ad-hoc research community group has proposed many multicast routing protocols, each one with its advantages and disadvantages in terms of robustness and efficiency to adapt to different network environment. For instance, the advantage of AMRIS protocol is that the nodes do not need to store any global information as it has a locality in link repair. By virtue of multicast updating cycle where each node broadcasts beacon messages containing its ID and other information, therefore, in case of a link failure, node with a bigger ID will rejoins the multicast tree rapidly. However, in case of rapid node moving and in term of robustness, AMRIS sees its performance decline, this is due to the fact that, it is classified as a tree-based multicast routing that doesn't have a redundant path between two nodes. Another drawback is when the node density increases, AMRIS needs to periodically send Beacon message to maintain the multicast tree, which will be greatly conflicted when the node's density is comparatively big. Despite AMRIS is relatively lightweight in terms of state, it is not in term of bandwidth, since control messages are sent periodically rather than being data-driven.

LAM is based on the CBT approach to building the multicast delivery tree, with one CORE to a group and provides multicasting service for large scale mobile ad hoc networks; LAM is not very robust, especially in a MANET environment. DLEMA is considered as a successful enhancement of LEMA especially in larger and denser WSN. It has the advantage of providing low delay and high success rate. However it is less robust in an environment which has a continuous link failure.

5. Conclusion

Mobile Health Care Training-based System (MHCTS) is still a relatively new field, where it can play an important role in the analysis of the collaboration between users, technology, and the healthcare domain. Such system involves the need of analysis to identify learning outcomes and test possible scenarios in 3D graphical environment before being applied to real life. MHCTS is a complex system that operates in complex networked environment to offer an immersive sense to all users in the VE. The goal of this paper was to identify the importance of multicast protocol in serving realistic aspect while they perceive user's actions distributed geographically. Another goal was to analyze some existing multicast protocols over ad-hoc network to identify the future direction of developing multicast protocol for Mobile Health Care Training-based System (MHCTS). Robustness and multicast efficiency are important factors to be considered in such application. If the degree of robustness is low, the packet delivery ratio will drop and high control overhead will be incurred. Also some other protocols implement the shortest path algorithm paths between senders and receivers. Thus, link failures and network latency are reduced. In this direction, as a future work, we will develop a new protocol that will take into consideration multicast efficiency and robustness in order to fill a gap in the E-health field.

Acknowledgements

Authors would like to express their thanks to Prince Salman Research and Translation Center (PSRTC) in Prince Sultan University.

References

[1]   Peterson, M. (2005) Learning Interaction in an Avatar-Based Virtual Environment: A Preliminary Study. *Journal Pacific Association for Computer Assisted Language Learning*, **1**, 29-40.

[2]   Paiva, P.V.F., Machado, L.S. and de Oliveira, J.C. (2012) A Peer-to-Peer Multicast Architecture for Supporting Collaborative Virtual Environments (CVEs) in Medicine. *Proceedings of 14th Symposium on Virtual and Augmented Reality*, Rio Janiero, 28-31 May 2012, 165-173.

[3]   Greenhalgh, C. and Benford, S. (1997) A Multicast Network Architecture for Large Scale Collaborative Virtual Environments. *ECMAST*97*, Milan, May 1997, 21-23.

[4]   Boukerche, A., Maamar, H. and Hossain, A. (2007) A Performance Evaluation of a Hybrid Multicast Transport Protocol for a Distributed Collaborative Virtual Simulation of a Brain Tumor Tele-Surgery Class of Applications. *12th IEEE Symposium on Computers and Communications, ISCC 2007*, Aveiro, 1-4 July 2007, 975-980.

[5]   Paiva, P.V.F., Machado, L.S. and Oliveira, J.C. (2012) An Experimental Study on CHVE's Performance Evaluation. *Studies in Health Technology and Informatics*, **173**, 328-330.

[6]   Boukerche, A. and Ren, Y. (2009) A Secure Mobile Healthcare System Using Trust-Based Multicast Scheme. *IEEE Journal on Selected Areas in Communications*, **27**, 387-399.

[7]   Claypool, M. and Claypool, K. (2006) Latency and Player Actions in Online Games. *Communications of the ACM, Special Issue: Entertainment Networking*, **49**, 40-45.

[8]   Wloka, M. (1995) Lag in Multiprocessor VR. *Presence: Teleoperators and Virtual Environments (MIT Press)*, **4**, 50-63.

[9]   Park, K. and Kenyon, V. (1999) Effects of Network Characteristics on Human Performance in a Collaborative Virtual Environment. *Proceedings of the IEEE Virtual Reality*, Houston, 13-17 March 1999, 104-111.

[10] Perkins, C.E. and Royer, E.M. (1999) Ad-Hoc On-Demand Distance Vector Routing. *Proceedings WMCSA'99. Second IEEE Workshop on Mobile Computing Systems and Applications*, 90-100. http://dx.doi.org/10.1109/MCSA.1999.749281

[11] Royer, E.M. and Perkins, C.E. (1999) Multicast Operation of the Ad-Hoc On-Demand Distance Vector Routing Protocol. In: *Proceedings of the 5th Annual ACM/IEEE International Conference on Mobile Computing and Networking*, ACM, New York, 207-218. http://dx.doi.org/10.1145/313451.313538

[12] Xiang, M. (2012) Analysis on Multicast Routing Protocols for Mobile Ad Hoc Networks. *Proceedings of the International Conference on Solid State Devices and Materials Science*, Macao, 1-2 April 2012, 1787-1793.

[13] Wu, C.W., Tay, Y.C. and Toh, C.K. (1998) Ad-Hoc Multicast Routing Protocol Utilizing Increasing Id-Numbers (AMRIS) Functional Specification. Internet Draft, November 1998.

[14] Garcia-Luna-Aceves, J.J. and Madruga, E.L. (1999) The Core-Assisted Mesh Protocol. *IEEE Journal on Selected Areas in Communications*, **17**, 1380-1394.

[15] Musznicki, B., Tomczak, M. and Zwierzykowski, P. (2012) Dijkstra-Based Localized Multicast Routing in Wireless Sensor Networks. *Proceedings of the 8th International Symposium on Communication Systems Networks & Digital Signal Processing (CSNDSP)*, Poznan, 18-20 July 2012, 1-6.

[16] Sanchez, J.A. and Ruiz, P.M. (2006) LEMA: Localized Energy-Efficient Multicast Algorithm Based on Geographic Routing. *Proceedings of the 31st IEEE Conference on Local Computer Networks*, Tampa, 14-16 November 2006, 3-12. http://dx.doi.org/10.1109/LCN.2006.322092

[17] Kruskal, J.B. (1956) On the Shortest Spanning Subtree of a Graph and the Traveling Salesman Problem. *Proceedings of the American Mathematical Society*, **7**, 48-50. http://dx.doi.org/10.1090/S0002-9939-1956-0078686-7

[18] Dijkstra, E.W. (1959) A Note on Two Problems in Connexion with Graphs. *Numerische Mathematik*, **1**, 269-271. http://dx.doi.org/10.1007/BF01386390

[19] Zhang, W., Jia, X., Huangand, C. and Yang, Y. (2005) Energy-Aware Location Aided Multicast Routing in Sensor Networks. *Proceedings of the International Conference on Wireless Communications, Networking and Mobile Computing*, Wuhan, 23-26 September 2005, 901-904.

[20] Garcia-Luna-Aceves, J.J. and Madruga, E.L. (1999) The Core-Assisted Mesh Protocol. *IEEE Journal on Selected Areas in Communications*, **17**, 1380-1394. http://dx.doi.org/10.1109/49.779921

[21] Ballardie, A., Francis, P. and Crowcroft, J. (1993) Core Based Trees (CBT): An Architecture for Scalable Inter-Domain Multicast Routing. *Proceedings of the ACM SIGCOMM' 93 Conference on Communications Architectures, Protocols and Applications*, San Francisco, September 13-17 1993, 13-17.

[22] Boukerche, A., Zarrad, A. and Araujo, R. (2007) A Novel Gnutella Application Layer Multicast Protocol for Collaborative Virtual Environments over Mobile Ad-Hoc Networks. *Proceedings of the IEEE Wireless Communications and Networking Conference*, Kowloon, 11-15 March 2007, 2825-2830.

8

Evaluation of Effective Vehicle Probe Information Delivery with Multiple Communication Methods

Tatsuya Yamada, Mayu Mitsukawa, Hideki Shimada, Kenya Sato

Mobility Research Center, Doshisha University, Kyoto, Japan
Email: ksato@mail.doshisha.ac.jp

Abstract

Vehicle probe information delivery systems can be broadly divided into the center type and center-less type. Since conventional center-type information delivery systems generate a large load on the communications infrastructure and data center, research efforts have come to be focused on the center-less type. However, existing vehicle probe information delivery systems suffer from various problems including a limited service area, low delivery efficiency, and lack of immediacy in delivery. Our objective in this study is efficient delivery of vehicle probe information as needed. We propose a delivery scheme that uses vehicle-to-vehicle communication, infrastructure-to-vehicle communication, and mobile communication as well as Geo cast. This combined use of multiple communication methods achieves efficient information delivery by changing the communication method to fit the current situation. The results of an evaluation by simulation showed that the proposed scheme could deliver information efficiently in a variety of environments.

Keywords

Vehicle Probe Information, ITS, Style, Vehicle-to-Vehicle Communication

1. Introduction

Research has been active in recent years on the delivery of vehicle probe information as one type of service using vehicle-to-vehicle (V2V) communication [1]. The methods used for delivering vehicle probe information can be broadly divided into the center type and center-less type. The center type uses either infrastructure-to-vehicle (I2V) communication or mobile communication: the former collects and provides vehicle probe information through roadside units while the latter does so through mobile terminals. The center-less type, meanwhile, uses

V2V communication. Exchanging information directly between vehicles is a method that excels in immediacy and makes infrastructure such as roadside units and base stations unnecessary. Since conventional center-type information delivery systems generate a large load on the communications infrastructure and data center, research efforts have come to be focused on center-less information delivery systems. Vehicle probe information collected and provided in one of the ways described above can be used to support a Driving Safety Support System, to alleviate congestion, etc. However, existing vehicle probe information delivery systems suffer from various problems including a limited service area, low delivery efficiency, and lack of immediacy in delivery. Our objective in this study is efficient delivery of vehicle probe information as needed, where efficient delivery is defined as the provision of information with high throughput and low delay. Specifically, we propose a delivery scheme that makes use of V2V communication, I2V communication, and mobile communication. This combined use of multiple communication methods achieves efficient information delivery by changing the communication method to fit the current situation.

2. Vehicle Probe Information Delivery Systems

2.1. System Configuration

A vehicle probe information delivery system treats a vehicle itself as a single moving sensor (referred to below as the "ego vehicle"). This system collects vehicle sensor information by various communication means and consolidates and processes that information to support a Driving Safety Support System, alleviate congestion, improve the environment, etc. Examples of services using a vehicle probe information delivery system include the delivery of hazardous location information based on driving behavior and optimal route guidance using congestion information.

2.2. Problems with Existing Delivery Systems

Vehicle probe information delivery systems can be broadly divided into the center type and center-less type. Infrastructure-to-vehicle communication employs Dedicated Short Range Communications (DSRC) [2]. In this type of communication, inter-vehicle interference cannot easily occur, which means reliable communications. However, the effective communication distance in DSRC is relatively short making for a limited service area. Mobile communication, meanwhile, uses wide-area radio communications enabling the collection and provision of information regardless of where the vehicle may be in an unlimited service area. On the other hand, the collection and provision of information in mobile communication is all achieved via transmissions with a data center, which detracts from immediacy.

In contrast to the above, V2V communication performs transmissions directly between vehicles without the need for base stations or other infrastructure facilities, which means a method that excels in immediacy. This method, however, collects and provides information only between vehicles. As a result, a small number of vehicles can make it difficult to collect and provide information while a large number of vehicles can generate congestion in the network resulting in a drop in delivery efficiency.

3. Proposed Scheme

3.1. Objective

In this study, we propose a delivery scheme that uses V2V communication, I2V communication, or mobile communication and Geocast communication as well. This combined use of multiple communication methods can achieve efficient delivery of probe information by switching the communication method as conditions change. Specifically, the scheme switches the method to be used by a vehicle according to the number of surrounding vehicles and to whether a roadside unit is present in the vicinity. The scheme will use V2V communication or I2V communication to deliver information that demands immediacy such as that for a Driving Safety Support System and will use mobile communication for supplementary information such as meteorological data. Vehicle-to-vehicle communication, in particular, will deliver information using Geocast communication to achieve efficient dissemination of information.

The target here is the delivery of traffic information and information that can be used to provides services to vehicles. A variety of applications can be envisioned including a Driving Safety Support System and the delivery of congestion reports, meteorological data, road conditions, and disaster information. With reference to materials [3] specifying delay-time requirements, we have established requirements for delay time with respect to a Driving Safety Support System that we envision to be the main application of our proposed scheme. Delay-time requirements differ depending on the support level.

3.2. Operation of Proposed Scheme

3.2.1. Switching of Communication Methods

The switching of communication methods in the proposed scheme depends on the number of vehicles surrounding the ego vehicle and the nearby presence of a roadside unit. The ego vehicle obtains the number of surrounding vehicles by using hello packets that it transmits by packet flooding limited to one hop. This enables the ego vehicle to count the number of vehicles with which a connection could be achieved and to then decide on the communication method to be used. Specifically, if the number of surrounding vehicles is equal to or greater than n1 and less than n2, the ego vehicle uses V2V communication, and if less than n1 or equal to or greater than n2, it uses mobile communication. In addition, the ego vehicle determines whether a roadside unit is nearby by using location information on the roadside unit and location information on itself. Location information on the roadside unit is obtained by referring to data recorded on a digital map within a vehicular application. The ego vehicle can then compare this data with its own location information obtained from the Global Positioning System (GPS) to obtain its distance from the roadside unit. If the roadside unit exists within a radius of Lm from the vehicle, the ego vehicle uses I2V communication.

Furthermore, in the proposed scheme, if peripheral conditions (number of surrounding vehicles and presence/absence of a nearby roadside unit) should change, the communication method will change in real time to match current conditions. Throughput and delay time in this scheme will therefore differ compared with delivery when using only a single communication method.

The operation of the proposed scheme is shown in **Figure 1**. In the figure, L denotes the distance between the vehicle and roadside unit and n1 and n2 denote the number of surrounding vehicles. The layered structure of the proposed scheme is as follows. First, on the application layer, the proposed scheme runs applications for V2V communication. These applications generate transmit data based on data obtained from vehicular devices, manage that transmit data, and determine the transmit area. Next, on the network layer, the proposed scheme transmits that data using Geocast or unicast communication. Finally, on the physical layer, the proposed scheme uses a network interface according to the communication method being used.

3.2.2. Use of V2V Communication

Delivery by V2V communication is used in an environment in which the amount of vehicular traffic is high so that a sufficient amount of information can be exchanged among vehicles. Specifically, V2V communication is used if the number of surrounding vehicles is equal to or greater than n1 and less than n2. In this case, information delivery to another vehicle is performed through Geocast [4] communication. A Geocast refers to a form of communication that uses location information instead of node IDs when transmitting data in an ad hoc network. In general, important factors in establishing communications between two nodes in an ad hoc network are the destination node ID and transmission path. In contrast, a key feature of Geocast communication is that there is no need to specify node IDs as is necessary in ordinary multicast communications, which means that there is no

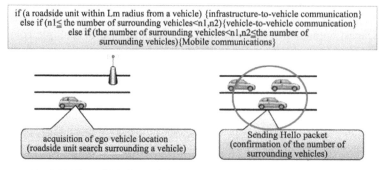

Figure 1. Proposed method.

need to learn the IDs of the destination nodes. In multicast communication, the packet header holds the destination node ID, while in Geocast communication, the packet header holds the ego vehicle's location information obtained using GPS together with information on the destination area. In this way, V2V communication transmits data to nodes in the destination area with individual vehicles performing relay transmission up to the destination area.

For relay transmission, the proposed scheme sets the destination area (Geocast area) and the relay area (forwarding area) from the location information specifying the destination. The setting of each of these areas is performed by vehicular applications. The Geocast area is set as a fixed area based on destination location coordinates. The forwarding area is set within the range of traffic lanes in both directions based on current location information of the information source and on road information. Here, road information is obtained from a map system that can detect the ego vehicle's position on the road from current location information. The forwarding area is set so as to encompass the information source and the Geocast area. Furthermore, in transmission by the proposed scheme, the following types of data are added to the transmit data: destination location information (destination coordinates representing a geographical location), ego-vehicle location information (information obtained from GPS), information set by vehicular applications (information indicating the range of the Geocast area and forwarding area), and time to live (TTL) information (to limit the range of flooding).

In relay transmission by the proposed scheme, the information source performs data flooding. A nearby vehicle that receives that data observes that the TTL of that packet is set to 1, which indicates that no re-flooding is to be performed. Next, this vehicle determines whether it belongs to the forwarding area based on location information obtained from GPS and forwarding-area information in the packet. If it does not belong, the data is destroyed. However, if the vehicle does belong to the forwarding area, it uses information on the destination area to transmit the data to the vehicle closest to the destination area among the vehicles situated within its transmission range. The data can therefore be relayed to the Geocast area by repeating this transmission process. Furthermore, if the ego vehicle should determine that it belongs to the Geocast area, it performs flooding to deliver the data to the vehicles in the destination area. When an ego vehicle sets out to transmit data to the vehicle closest to the destination area among vehicles within its transmission range, it limits candidate vehicles to those within the forwarding area and determines the location of nearby vehicles by exchanging hello packets with them. In this way, the ego vehicle determines its positional relationship with nearby vehicles enabling it to decide which of those vehicles will be its transmission destination.

Relaying data using only vehicles within the forwarding area in this way reduces the communications load on the network. In addition, including a vehicle's orientation information in vehicle probe information can further limit the delivery area. For example, orientation information can be used to propagate information on a sudden stop only to vehicles behind the ego vehicle thereby keeping the amount of communication needed to a bare minimum.

3.2.3. Use of I2V Communication

Delivery by I2V communication is used in an environment having a high vehicle density and including, for example, a traffic intersection at which a roadside unit is installed. In particular, I2V communication is used if a roadside unit exists within a distance of Lm from the vehicle. Location information on the roadside unit is obtained by referring to data recorded on a digital map within a vehicular application. An ego vehicle obtains its own location information and compares it with that of the roadside unit to decide whether to use I2V communication.

3.2.4. Use of Mobile Communication

Delivery by mobile communication is used in an environment in which the amount of vehicular traffic is low making the passing of information to another vehicle difficult, or in an environment in which vehicle density is so high so that V2V communication would generate congestion in the network. Specifically, mobile communication is used if the number of surrounding vehicles is less than $n1$ or equal to or greater than $n2$. In general, mobile communication is used to deliver information that does not demand immediacy compared to other types of information. Mobile communication uses the 3G network and can therefore deliver information within a service area that is essentially unlimited. It also enables the provision of information using pull-type communication depending on the needs of the driver.

4. Performance Evaluation by Simulation

4.1. Simulator

In this study, we used the Qualnet [5] simulator to evaluate the performance of the proposed scheme. The parameters used in this simulation are listed in **Table 1**. For packet size, we adopted the value generally used in papers related to V2V communication [6]-[8]. Furthermore, with the aim of replicating communications in an actual environment, we used TWO-RAY as the spatial model since it takes ground-reflected waves into account [6]. Additionally, we decided on antenna height and radio output range by referring to transmission antennas mounted on ordinary vehicles [7]. Finally, for the MAC layer and packet size, we adopted the system and value generally used in V2V communication and having a proven track record in research and development [6]-[8].

4.2. Evaluation Models

To compare the proposed scheme and existing schemes in terms of delivery efficiency and delivery immediacy, we measured throughput (amount of data transmitted per second) and delay time (time taken for data transmitted from the information source to arrive at its destination) for various evaluation models as described below.

4.2.1. Evaluation Model 1

For evaluation model 1, we simulated delivery by various communication systems that can be used for V2V communication with the aim of demonstrating the usefulness of Geocast as used in V2V communication by the proposed scheme. To test throughput and delay time under conditions in which V2V communication can generally be used, we assumed an environment in which groups of vehicles pass each other within an area having an intersection. Given this environment and referring to Advanced Safety Vehicle (ASV) study materials [9] of the Japan Automobile Manufacturers Association (JAMA), we set driving speed to 45 km/h, inter-vehicle distance to 20 m, and number of vehicles to 10 - 20. We assumed that transmit data would consist of approaching-vehicle information and set its size to 2 MB including graphics and text data, which is basic to vehicle probe information. The specific targets of comparison that we used for this evaluation were communication systems commonly used for V2V communication, namely, flooding (with TTL limitations set so that no retransmission occurs when the same data is received multiple times), Optimize Link State Routing (OLSR), Ad hoc on-Demand Distance Vector (AODV), and Geocast. Evaluation model 1 is outlined in **Figure 2**.

4.2.2. Evaluation Model 2

For evaluation model 2, we simulated delivery when changing parameters (n1, n2) in the proposed scheme with the aim of deriving the optimal values for those parameters. Here, to assess throughput and downlink for various combinations of n1 and n2 in the delivery of vehicle probe information in an actual environment, we referred to ITS Communication Simulation Evaluation Scenarios [9] and performed the simulation for an environment in which some vehicles were traveling in the east-west direction and other vehicles were stationary in the north-south direction within a 2.5 × 3.5 km area. Furthermore, referring again to ASV study materials of JAMA, we set driving speed to 60 km/h, inter-vehicle distance to 45 m, and number of vehicles acting as information sources to 10 - 40. Transmit data consisted of 3 MB of sensor data (graphics, etc.). In the simulation, we measured average throughput and average delay time for 10 - 40 vehicles acting as information sources. We first varied n2 while setting n1 constant to determine an optimal value for n2, and after determining n2, we varied n1 to determine its optimal value. Specifically, for n2, we adopted the value that recorded highest throughput and lowest delay time. Evaluation model 2 is outlined in **Figure 3**.

Table 1. Simulation parameters.

Simulator	Qualnet ver. 5.0.1
Spatial model	TWO-RAY: consideration of ground-reflected waves
Antenna height	1.5 m
MAC layer	IEEE802.11b
Packet size	512 byte
Radio output range (effective radius)	V2V communication: 50 m I2V communication: 25 m

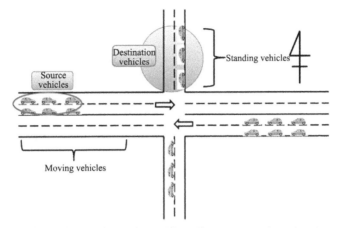

Figure 2. Outline of evaluation model 1.

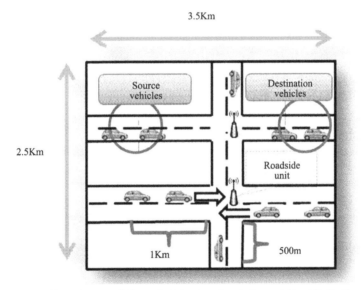

Figure 3. Outline of evaluation model 2.

4.2.3. Evaluation Model 3

We used evaluation model 3 to assess throughput and delay time for various techniques assuming the delivery of vehicle probe information in an actual environment. Specifically, we performed a simulation for an environment in which vehicles were traveling in the east-west direction within a 2.5 × 3.5 km area. The simulation model that we used for this evaluation was the same as evaluation model 2. Referring again to ASV study materials, we set driving speed to 60 km/h, inter-vehicle distance to 45 m, and number of vehicles acting as information sources to 10 - 40. Transmit data consisted of 3 MB of sensor data (graphics, etc.). The communication methods that we compared in this case were V2V communication (AODV), I2V communication (amplitude-shift keying (ASK)), mobile communications (3G), and the proposed scheme. For each of these communication methods, we adopted a communication system that is typically used for that method [10].

4.3. Simulation Results

4.3.1. Evaluation Model 1

For evaluation model 1, we measured throughput and delay time for various communication systems used in V2V communication. It was found that delivery using Geocast achieved high throughput and low delay by transmitting information only within a specific range. Simulation results for evaluation model 1 are shown in **Figure 4** and **Figure 5**.

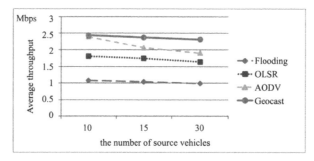

Figure 4. Results for evaluation model 1 (throughput).

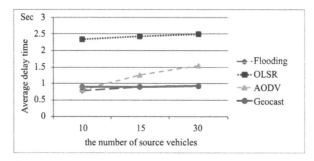

Figure 5. Results for evaluation model 1 (delay time).

4.3.2. Evaluation Model 2

For evaluation model 1, we measured throughput and delay time while varying parameters (n1, n2) in the proposed scheme. The results showed that high-throughput and low-delay communications could be achieved for n1 = 2 and n2 = 6, so these values were taken to be optimal for those parameters. Simulation results for evaluation model 2 are shown in **Figure 6** and **Figure 7**. Specifically, for n2, we varied its value in the manner of 2, 3, 4, 5,... and adopted the value for which the highest throughput and lowest delay time were measured.

4.3.3. Evaluation Model 3

For evaluation model 3, we measured throughput and delay time for V2V communication (AODV), I2V communication (ASK), mobile communications (3G), and the proposed scheme. Results showed that the proposed scheme maintained high throughput regardless of the number of source vehicles since it was able to perform information delivery by changing the communication method as needed. The parameters used in the simulation of the proposed scheme were L = 50, n1 = 2, and n2 = 6. Simulation results for evaluation model 3 are shown in **Figure 8** and **Figure 9**.

5. Discussion

5.1. Evaluation Model 1

Geocast communication reduces wasteful transmission by transmitting information only to those vehicles needing it in a limited area. This reduces the load on the network resulting in high throughput. Flooding, on the other hand, transmits data to all vehicles, which increases the load on the network and causes many packets to drop thereby reducing throughput. In OLSR, meanwhile, vehicles are always moving in an east-west direction within the communication area resulting in significant changes in the routing table. As a result, transmissions increase to update the routing table and throughput drops. AODV, in contrast, is a scheme that does not continuously maintain a routing table, so a route must be created for each destination. This means that transmissions involved in route control increase as the number of vehicles increase, which causes throughput to drop. Both Geocast and flooding require a small amount of control when transmitting data to multiple vehicles, and as a result, delay time is small. In OLSR, the routing table changes greatly when vehicles moving toward each other pass each other. Transmissions are therefore needed at this time to update the routing table, which causes delay time to increase. In AODV, the time involved in creating routes depends on the number of vehicles, so time is needed to

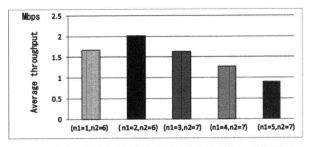

Figure 6. Results for evaluation model 2 (throughput: n2 is optimal number).

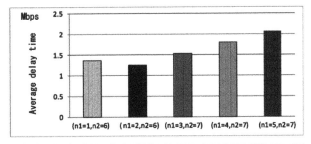

Figure 7. Results for evaluation model 2 (delay time: n2 is optimal number).

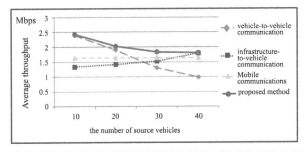

Figure 8. Results for evaluation model 3 (throughput).

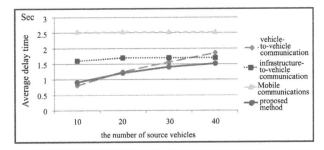

Figure 9. Results for evaluation model 3 (delay time).

establish routes in an environment with a high density of vehicles thereby increasing delay time. Based on the above discussion, delivery by Geocast is superior in terms of throughput and delay time.

5.2. Evaluation Model 2

The results obtained from simulations while varying n1 and n2 show that throughput would be low and delay time would be long if the value of n1 is high or the value of n2 is low since mobile communication would be used in such cases even if other vehicles are near the ego vehicle. Furthermore, for n2 set above a certain level (n2 = 6, 7), it was found that throughput would be low and delay time would be long owing to network congestion since V2V communication would be used regardless of whether other vehicles are concentrated about the ego vehicle. Accordingly, optimal conditions for switching communication methods are the presence of other vehicles at a number that does not generate congestion in the network. Based on the results of these simulations, we concluded that these optimal conditions occur for n1 = 2 and n2 = 6.

5.3. Evaluation Model 3

In V2V communication, communication load on the network increases as the number of vehicles that act as information sources increase thereby degrading throughput. In I2V communication, an increase in the number of vehicles means that more vehicles will come to lie within the range of communication with the roadside unit, which improves average throughput. Mobile communication, meanwhile, maintains constant throughput since it uses 3G circuits. The proposed scheme can maintain high throughput by switching among these communication methods to deliver information. In addition, V2V communication excels in immediacy since it enables vehicles to exchange data directly among themselves without having to use a data center. Here, however, delay time increases as the number of vehicles that act as information sources increase. Meanwhile, in I2V communication, delay time increases if vehicles lie outside the range of communication with the roadside unit since regular transmissions cannot be performed. In mobile communication, delay time is long since transmissions must pass through a data center. The proposed scheme uses V2V communication as its main communication method, so average delay time is short. In addition, these simulations showed that high throughput and low delay could be achieved for parameters set to L = 50, n1 = 2, and n2 = 6. In short, we have shown that the proposed scheme is superior in terms of throughput and delay time compared with existing delivery schemes.

6. Related Research

Takahata *et al.* proposed a vehicle probe information delivery system using I2V communication [11]. They developed, in particular, a center-type probe information delivery system using DSRC. This system, however, can only collect and provide information in areas within the vicinity of roadside units, so it is not oriented to generating local information. Our proposed scheme, in contrast, can provide services regardless of the environment. Ito *et al.* proposed a vehicle probe information delivery system using V2V communication [12]. In this system, information is exchanged only between vehicles, and as a result, service quality is somewhat affected by the number of vehicles present. Our proposed scheme, meanwhile, uses multiple communication methods and is consequently unaffected by the number of vehicles, so it excels in terms of service provision.

Ushitani *et al.* also proposed a vehicle probe information delivery system using multiple communication methods [13]. It can collect vehicle probe information over a wide area through a communication network that combines V2V and I2V communication methods. This system looks to be applicable to traffic accident prevention, hazard avoidance, and congestion alleviation. However, the basis for communication in this system is the vehicle, and as a result, communications cannot be established in an environment having only a few vehicles. That is to say, the system assumes an area with a high concentration of vehicles even in an applicable environment—using it in an area with only a scattering of vehicles is difficult. Our proposed scheme, though, can use mobile communication even in an area with a scarcity of vehicles, so it excels in the ability to collect and provide vehicle probe information.

Finally, Komiya *et al.* developed an evaluation platform using a center-less type of information delivery system [14]. Their research showed that center-less type of probe information delivery is superior in terms of a CO_2 reduction effect. We can therefore infer that our proposed scheme is a feasible application.

7. Conclusions

In this study, we proposed a delivery scheme that uses vehicle-to-vehicle communication, road-to-vehicle communication, or mobile communication to enable efficient delivery of large volumes of information without limiting the service area. Combining multiple communication methods in this way enables the communication method to be changed depending on current conditions so that information can be delivered in the most efficient way.

In simulations that we performed to evaluate the proposed scheme, we measured throughput and delay time for both the proposed scheme and existing schemes in envisioned environments. Simulation results revealed that switching communication methods by the proposed scheme maintained high throughput and low delay and achieved efficient delivery of data. These results demonstrate that the proposed scheme is superior to existing schemes in terms of delivery efficiency and delivery immediacy.

Acknowledgements

This research was supported partly by a Grant-in-Aid for Scientific Research (24300030) and the Strategic Information and Communications R&D Promotion Programme (12180615) by the Ministry of Internal Affairs and Communications, Japan.

References

[1] Japan Automobile Research Institute. http://www.jari.or.jp/

[2] Kenney, J.B. (2011) Dedicated Short-Range Communications (DSRC) Standards in the United States. *Proceedings of the IEEE*, **99**, 1162-1182. http://dx.doi.org/10.1109/JPROC.2011.2132790

[3] ETSI (2013) Intelligent Transport Systems V2X Application Road Hazard Signalling Application Requirements Specification. http://www.etsi.org/deliver/etsi_ts/101500_101599/10153901/01.01.01_60/ts_10153901v010101p.pdf

[4] Navas, J.C. and Imielinski, T. (1997) GeoCast—Geographic Addressing and Routing. *Proceedings of International Conference on Mobile Computing and Networking (MobiCom)*, Budapest, 26-30 September 1997, 66-76.

[5] Scalable Network Technologies. http://www.scalable-networks.com/

[6] Haerri, J., Filali, F. and Bonnet, C. (2006) Performance Comparison of AODV and OLSR in VANETs Urban Environments under Realistic Mobility Patterns. *Proceedings of the 5th IFIP Med-Hoc-Net'06*, Lipari, June 2006, 3266-3277.

[7] Harri, J., Filali, F. and Bonnet, C. (2006) Mobility Models for Vehicular Ad Hoc Networks: A Survey and Taxonomy, Technical Report RR-06-168, Institute Eurecom, Sophia Antipolis.

[8] Zhao, J., Zhang, Y. and Cao, G.H. (2007) Data Pouring and Buffering on the Road: A New Data Dissemination Paradigm for Vehicular Ad Hoc Networks. *IEEE Transactions on Vehicular Technology*, **56**, 3266-3277. http://dx.doi.org/10.1109/TVT.2007.906412

[9] Japan Automobile Research Institute, Committee Promoting the Use of ITS Simulators, ITS Communication Simulation Evaluation Scenarios, 2012. (In Japanese) http://www.jari.or.jp/Portals/0/resource/pdf/H23_simyu/%EF%BC%88Ver1.2%EF%BC%8920131010.pdf

[10] Japan Automobile Research Institute, ITS Research Division, Report on Feasibility Study on Development of Center-Less Probe Information System, 2007.

[11] Takahata, K., Noda, T. and Kamba, K. (2009) Information Provision System Using DSRC (Spot Communications). http://www.hrr.mlit.go.jp/library/happyoukai/h22/ino_2/209.pdf

[12] Ito, H. (2011) Developing Decentralized Probe Information Systems Utilizing 700MHz Band for Mobile Communication and Advertizing Activity on Its Achievements. *JARI Research Journal*, **33**, 55-58.

[13] Ushitani, Y., Imao, M., Higashino, T., Tsukamoto, K. and Komaki, S. (2006) Improvement Effect in Link Blocking Rate for Joint Inter-Vehicle and Road-to-Vehicle Communication System with Large-Sized Vehicle Operation. *B-Abstracts of IEICE TRANSACTIONS on Communications*, **J89-B**, 909-919.

[14] Komiya, T., Horiguchi, R. and Koide, K. (2011) Estimation of CO_2 Reduction with the "Center-Less" Probe System. *Proceedings of the 10th ITS Symposium*. (In Japanese) http://www.transport.iis.u-tokyo.ac.jp/publication/2011-32.pdf

9

Comparative Study of Proactive, Reactive and Geographical MANET Routing Protocols

Muthana Najim Abdulleh[1], Salman Yussof[1], Hothefa Shaker Jassim[2]

[1]College of Information Technology, Universiti Tenaga Nasional, Kajang, Malaysia
[2]College of Engineering, Komar University of Science and Technology, Sulaymaniyah, Iraq
Email: mut.n707@yahoo.com, salman@uniten.edu.my, hothefa.shaker@komar.edu.iq

Abstract

Mobile Ad-hoc Network (MANET) is defined as a combination of mobile nodes that lack a fixed infrastructure and is quickly deployable under any circumstances. These nodes have self-aware architecture and are able to move in multiple directions, which renders it dynamic topology. Its dynamicity makes routing in MANET rather challenging compared to fixed wired networks. This paper aims to perform a comparative study on the three categories of MANET routing protocol by comparing their characteristics and operations, as well as their strength and weaknesses.

Keywords

MANET, Proactive Routing, Reactive Routing, Geographical Routing

1. Introduction

Mobile Ad-hoc Network (MANET) is defined as a combination of mobile nodes that can keep in touch with one another despite the lack of a core centralized administrator or fixed infrastructure [1]. MANET possesses a dynamic temporary network topology, which makes it rapidly deployable in a situation whereby setting the wired network is almost impossible, such as in the battlefields and natural disaster areas. Due to its dynamic topology, wireless nodes in MANET function both as a host and a router to keep the internal communications network active. The wireless nodes within MANET move in an arbitrary fashion and organize themselves in a random manner. Direct communication between wireless nodes happens if they are within the range of radio transmission. If not, communication is established via intermediate nodes, which forward packets and recognize MANET as a multi-hop network [2]. MANET also has a lower bandwidth than that of wired network and since it is operating on batteries, its operation must be energy efficient to maximize the life span of the nodes [3].

Over the years, there are many different routing protocols that have been developed for MANET. In general, these protocols can be categorized into three types: proactive, reactive and geographical routing protocols. This paper presents a comparative study of these three categories of MANET routing protocols. The presentation of the paper is organized as follows. Section 2 classifies multiple MANET routing protocols and provides a brief overview of several protocols in each category. Section 3 presents a comparison between the MANET routing protocols. An analysis of MANET routing protocols in terms of their characteristics, operation, strengths and weaknesses is presented in this section. It also highlights the drawbacks of these routing protocols to identify the areas that can be improved. Section 4 concludes the present comparative study of routing protocols in mobile ad-hoc networks.

2. Classification of Routing Protocols in MANET

In MANET, routing protocols can be classified into Proactive Routing Protocols, Reactive Routing Protocols, and Geographical Routing Protocols [4] [5]. **Figure 1** displays the basic classification of the routing protocols for MANET.

2.1. Proactive Routing Protocols

Proactive or table-driven routing protocols aim to keep up-to-date routing information flowing throughout a network between all nodes. As a means of preserving a consistent network connection, proactive routing protocols require every node to support at least one table which contains routing information. These nodes then react to the variations in the topology of the network by distributing the most current information through the network. This type of protocols is unique compared to others in term of the manner which the alterations to the network's structure are transmitted and also the amount of routing-related tables that are required. The benefit of proactive routing protocols is the median delay time per packet that can be decreased. In these protocols, route information is present and accessible in the table whenever it is required. Nevertheless, in maintaining up-to-date routing information, proactive protocols uninterruptedly employ a significant share of network capacity. This makes such routing protocols unsuitable for reconfigurable mobile ad-hoc networks [6] [7]. Additionally, further network capacity wastage occurs as the majority of stored routing information actually may never be used, and the node activity is fast and the variations in topology are more regular than the actual requests for route information. A summary of the variety of proactive routing protocols will be given in subsequent sections.

2.1.1. Optimized Link State Routing Protocol (OLSR)
The OLSR is an optimized pure link state algorithm with proactive nature which allows it to ensure the availability of the routes when required. Hop-by-hop mechanism is utilized to forward packets, which is one of the main characteristics of any MANET routing protocol [6] [7]. Being a link-state routing algorithm, OLSR needs to keep up-to-date information about the nodes in the network and the route to each of these nodes. This is done by having the nodes to periodically broadcast link-state update messages. These updates may cause a large

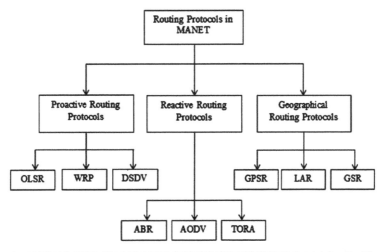

Figure 1. Classification of routing protocols.

amount of traffic to be generated, which may take up network resources and reduce the performance of the network. Multi Point Rely (MPR) is a unique feature of MPR that is able to minimize the number of rebroadcasting nodes and, in turn, reduce the number of control messages generated during an update. With MPR, nodes are able to exchange topological information in a periodical manner without having to generate a large amount of traffic [6]. **Figure 2** shows how update messages are transmitted with the use of MPR. Notice that the updated messages are only forwarded via selected nodes [8]. A route, from the given source to the destination is also created by the MPR. The process of neighbor detection is carried out by the periodical broadcast of HELLO messages by each node that are linked. These nodes sense each other, and in cases of symmetrical links, will regard each other as neighbors. Furthermore, link sensing and MPR selection process can also be carried out by the HELLO messages. All information pertaining to the relevant node that sent the HELLO message and its neighboring node will be found in the HELLO message. Each node is updated and recalculated when the updated information is received [9]. The TC message broadcasts topological information throughout the network, but these messages are only forwarded through MPR nodes. The MPR strategy functions better in large networks, which also positions OLSR as a better choice for large and dense networks.

1-and-1 hop symmetrical information is used by the MPR selection process to recalculate the MPR set. When a change in 1 or 2 hop neighborhood's topology is detected, then the MPR recalculation will occur. The route to each known destination is recalculated and updated when the updated information is received [10]. MPR functions better in larger networks, which means that OLSR is better suited for large and dense networks [11].

2.1.2. Destination-Sequenced Distance-Vector (DSDV)
Perkins and Bhagwat introduced Destination-Sequenced Distance-Vector (DSDV) [12], which is regarded as the premier ad hoc routing protocols, similar to many distance-vector routing protocols [13]. It was expanded from the classical Bellman-Ford routing mechanism [14]-[16], with the inclusion of destination address, sequence number and the number of hops rendering which it suitable for MANET operations. Each node possesses a routing table equipped with one route entrance to each goal, where the shortest path route (based on the number of hops) is recorded. A destination sequence number is utilized for the purpose of eschewing routing loops. A node increases its sequence whenever an alteration occurs in its vicinity. To maintain routing table consistency, routing changes are constantly passed throughout the network. There are two types of updates that are used; full dump and incremental. A full dump passes the whole routing table to its neighbors and is capable of acquiring many network protocol data units (NPDUs). Incremental updates are minuscule (must fit in a single packet) and transmits entries from routing tables that are altered during the previous full dump update. When the network is stabilized, updates are passed on, with full dumps seldom occurring. However, full dumps are common in a quick paced network. On top of the information being provided by the routing table, every route update packet possesses a unique sequence number that was assigned by the transmitter. The routes being labelled are frequently updated (highest number) using sequence number. The shortest route will be automatically selected if

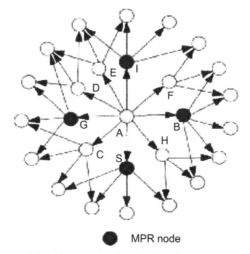

Figure 2. OLSR multipoint relay shows how update messages are transmitted with the use of MPR [11].

any two routes share a similar sequence number [13]. When there are multiple routes to a destination, a node will choose a route with the highest destination sequence number. This guarantees the utilization of routes that contains updated information [17]. Updated route broadcast will post the address of the destination, the number of hops to reach the destination, the sequence number of the destination and a new unique sequence number to broadcast; a route that is assigned an updated sequence number is regarded as a new route. However, if the sequence numbers are discovered to be similar, then a superior metric will be chose [14].

2.1.3. Wireless Routing Protocol (WRP)

WRP is regarded as a part of the general class of path-finding algorithms [12] [18] [19], and defined as a collection of distributed shortest path algorithms that determines the paths via information on the length and second-to-last hop of the shortest path to every destination. WRP decreases the number of cases whereby a temporary routing loop might occur. Wireless Routing Protocol is a table-based protocol with the intention of keeping routing information secured between the nodes of the network. Every node is tasked with maintaining four tables [3] [15] [20] [21]:

• Distance table
• Routing table
• Link-cost table
• Message retransmission list (MRL) table

WRP utilizes periodic update message transmissions adjacent to a node. The nodes in the response list of new messages (which is formed using MRL) should in turn acknowledge it. If there are no changes from the previous update, then the nodes in the response list will send an idle HELLO message to confirm connectivity. A node is empowered to decide whether or not to update its routing table post-receiving an updated message from nearby while looking for a superior path with the updated information it receives. In case the node obtains a superior path, this information will be relayed to the original node for table updates. After being acknowledged, the original node will proceed to update its MRL. Every time the consistency of the routing information is being examined by the nodes present in the protocol, it helps reduce routing loops and determine the best routing solution within the network [22].

2.2. Reactive Routing Protocols

Reactive routing is also known as on-demand routing protocols. These protocols lack routing information or routing activity on the nodes in the network when communication is lacking or dismal. Unused routes are maintained with less overhead. Unfortunately, more time delays may be experienced initially. A route is searched for by the reactive protocol in an on-demand manner if a node intends to pass on a packet to another node. The packet is received and transmitted after forming a connection. Then, the request packets are dispersed within the networks, leading to a discovery of routes. There are two categories of reactive protocol; source routing and hop-by-hop routing. A complete source to the designated address is carried by the source routed on-demand protocols. The information contained in the header of each packet will be evaluated by every intermediate node when forwarding these packets. The intermediate nodes are not required to maintain updated routing data for each active route. In addition, neighbor connectivity through periodic beaconing messages is also not required in the database of the nodes. As each node has the potential to update its routing table in the presence of fresher topology information, the routes are therefore adaptable to the changing environment, which takes place dynamically in the MANETs. The data packets are forwarded over better and fresher routes this way [23]-[25]. The protocols under this category are detailed below.

2.2.1. Associativity-Based Routing (ABR)

The ABR [26] [27] protocol is responsible for detailing a unique routing metric called "degree of association stability" for mobile ad-hoc networks. It is assumed to lack loops, deadlock, and packet duplicates. In ABR, routes are chosen based on the links between the states of nodes. The chosen nodes are expected to last long enough. Each node gives off periodic beacons as a mark of its existence. When an adjacent node detects a beacon, it will promptly update its associated tables. Each beacon received will result in an increment to the node. In this case, association stability also means the connection stability of a node towards another at the same time and space. A higher value of associativity ticks with a node indicates a low state of node mobility, and vice versa.

Associativity ticks revert to normal when the neighbors or the nodes move beyond their respective reach. The fundamental objective of ABR is to seek out longer-lived routes for ad hoc mobile networks, which are Route discovery, Route reconstruction (RRC), and Route deletion [3] [28].

1) Route discovery phase: The route discovery phase is a broadcast query and await-reply (BQ-REPLY) cycle. The source node broadcasts a BQ message seeking nodes possessing a route to the destination. A node will only pass a BQ request once. Upon receiving the BQ message, an intermediate node alters both the address and associativity ticks of the query packet. The upcoming node will delete the upstream nodes of its neighbors' associativity tick entries while keeping the entry that is associated with itself and its corresponding upstream node. Every packet that reaches the destination will possess the associativity ticks of the nodes along the route, all the way from the source to the destination. Now, the destination can freely choose the best route via analysing the associativity ticks associated with each path. In the case of multiple paths possessing similar degree of association stability, the route having the least amount of hops will be chosen. Once a path has been determined, the destination passes forth a REPLY packet back to the source on this path. The nodes that the REPLY packet adhere to will serve to validate their respective routes, while other routes remain inactive, eschewing any chances of duplicated packets reaching the destination as well [28].

2) Route reconstruction (RRC) phase: RRC phase consists of partial route discovery, invalid route erasure, valid route updates, and new route discovery, depending on which node(s) along the route move. The movement of source nodes will precipitate a unique BQ-REPLY process due to the fact that the routing protocol is source-initiated. The route notification (RN) message deletes entries associated with routes and downstream nodes. When the destination moves, its immediate upstream node deletes its corresponding routes. A localized query (LQ [H]) process, where H refers to the hop count from the upstream node to the destination, will start for the purpose of confirming whether or not the node can be reached. If the destination gets the LQ packet, it will be prompted to choose the best partial route and REPLYs; otherwise, the initiating node times out and backtrack to the next upstream node. An RN message is dispatched to the adjacent upstream node to delete invalid routes and also inform it that the node must initiate the LQ [H] process. However, if the backtracking exceeds halfway to the source, the LQ process is terminated, and the source will restart the BQ process all over again [28].

3) Route deletion phase: When a route is no longer required, the source node will start a route delete (RD) broadcast. Each node present on the route will remove the route's entry from their respective routing tables. The RD message is broadcasted indirectly, as the source node might be unaware of any alteration to its route during RRCs [28].

2.2.2. Temporally Ordered Routing Algorithm (TORA)

TORA is an adaptive routing protocol for highly dynamic mobile multi hop networks that are source initiated and based on link reversal algorithms [29]. This protocol is able to rapidly build routes and reduce communication overhead via the localization response to topological alterations as much as they can [30]. TORA uses the "direction of the next destination" to send data, instead of using the concept of the shortest path to determine routes. This means less processing and less bandwidth usage. The source node uses one or two paths to the destination through several intermediate neighboring nodes [31]. The three main processes in the TORA protocol are route creation, route maintenance and route erasure. The route creation process uses query and UDP packets. For route creation, a height metric is used, where the height of the destination node is set to 0, while all of the others are set to NULL. The source node will then proceed to transmit a query packet containing the destination node's ID. Nodes that possess a non-NULL height will respond using a UDP packet that is made up of its height. The node receiving the UDP packet is set at a height that's higher, and is regarded as being "upstream" and vice versa. This results in the construction of a direct acyclic graph (DAG), from source to destination. The route formation process is realized by sending a request from the source, and receiving replies from its intended destination. During mobility, the DAG is broken, and route maintenance will then work to restore a DAG that is routed at the destination [32].

2.2.3. Ad-Hoc on Demand Distance Vector Routing Protocol (AODV)

Ad-hoc On-Demand Distance Vector Routing Protocol (AODV) [33]-[36] is a unicast reactive routing protocol. Basically this implies that the routes are formed when they are needed. The AODV protocol contains four control packets; HELLO messages, route requests (RREQs), route replies (RREPs) and route error messages (RERRs). These control packets are used in the two protocol mechanisms which are route discovery and route maintenance. In the AODV protocol, all nodes maintain a routing table that stores information regarding active routes. The information stored are destination, next hop, number of hops, sequence number for the destination, active neighbors for a route and the expiration time for a route table entry. Route entry timeout are updated upon usage. To prevent looping in distance vector routing, a sequence number is sent with RREQs and RREPs, both of which are stored in the routing table. A larger sequence number is indicative of the fact that recent updated route information and the one with the highest sequence number will be utilized. If two routes possess the same sequence number, the one with the fewer number of hops (a shorter route) will be used.

Route discovery mechanism begins when no valid route is found within the routing table of the source node. Route requests (RREQs) are sent to the network to search for the route to the destination. Receiving nodes create reverse routing entries towards the source for the purpose of sending possible reply packets later. A route reply (RREP) is dispatched by either the destination or intermediate node that is a validated route towards the destination. Nodes that received RREPs also create reverse routing entries towards the nodes that sent the RREPs. Often, each of the nodes along an active route will transmit HELLO messages to the neighboring nodes. If no HELLO message or data is received from a neighboring node after a period of time, the link is regarded as broken. If the destination of the route using this link is nearby the next hop from the neighbor, then a local repair process may be used to repair the route. If not, then a route error (RERR) message is sent to neighboring nodes, which then broadcasts the RERR message towards other nodes that may have routes affected by the broken link. If the route is needed by the affected source, the route discovery process will then be repeated [36].

2.3. Geographical Routing Protocols

Geographical routing [37] [38] utilizes information derived from a location for the purpose of formulating and optimizing the searching route towards the destination. Geographical routing suits sensor networks, especially where data aggregation remains a useful technique in the minimization of transmission to the base station via the elimination of redundancy between packets from multiple sources [39]. There is also a higher possibility for big multi-hop wireless network topology to change frequently. Geographical routing needs only the propagation of single-hop topology information such as the optimal neighbor to decide accurately on forwarding. The way it localizes its approach decreases the requirement of maintaining the routing tables, which in turn decreases the control's overhead, and eliminated the need for flooding. The nodes that are within the marked forwarding zones are capable of forwarding data packets. This marked region can be defined by the source or intermediate nodes to exclude nodes that might precipitate a detour in the course of forwarding the data packet. The second property associated with geographical routing is position-based routing. In this case, a node only needs to know where its direct neighbor is located. The mechanism that is involved in this case is the greedy mechanism whereby each node forwards a packet to an adjacent node. The Euclidean distance to the destination will be utilized as a metric. Position-based routing protocols are capable of reducing the overhead and energy because flooding for node discovery and state propagation is localized within a single hop [39]. The network density, accurate localization of nodes, and the forwarding rule are the deciding factors for the efficiency of the scheme [40].

2.3.1. Greedy Perimeter Stateless Routing (GPSR)

Greedy Perimeter Stateless Routing (GPSR) is a novel routing protocol for wireless datagram networks that utilizes the location of the routers and its destination to decide on forwarding. GPSR decides on greedy forwarding decisions by utilizing the information regarding a router's adjacent neighbors within the network's topology. When a packet reaches a region within which greedy forwarding becomes impossible, the algorithm recovers itself via routing adjacent to the perimeter of the region. By remaining close to the local topology, the GPSR scales better in per-router state than shortest-path and ad-hoc routing protocols as the number of network destinations increases. Under the mobility's frequent topological changes, **Figure 3** shows that the GPSR can utilize local topology information to discover accurately new routes faster. The local topology information can be used by GPSR to search for the new routes immediately, even under constantly changing topology due to node mobility. When choosing a packet's next hop, a forwarding node can make a local greedy choice optimally. This is because the initiator has marked the packet with the location of its destination under GPSR. The GPSR protocol utilizes extensive simulation of mobile wireless networks for the purpose of comparing its performance with that of Dynamic Source Routing [41].

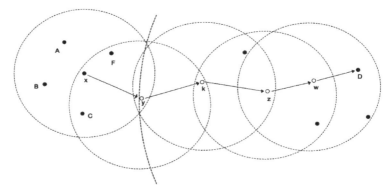

Figure 3. Greedy perimeter stateless routing [41].

2.3.2. Location Aided Routing (LAR)

The location aided routing protocol fails to confirm a location-based routing protocol and proposed the usage of position information to improve the route discovery phase of reactive ad-hoc routing approaches. The location information is obtained by GPS by utilizing two flooding regions; forwarded and expected. The decrease in the search space will inevitably result in lesser route messages. When a source node intends to dispatch a data packet, it will first request for the location of the destination from the location service which causes contacting and tracking problems [42]. **Figure 4** as shown in the two LAR algorithms being proposed; LAR **Scheme 1** and LAR **Scheme 2** [43]. LAR **Scheme 1** utilizes the expected location of the destination to confirm the requested zone during route discovery. The requested zone is rectangular both at the source and at the expected zone of destination. Its sides are parallel to both the x and y-axis. In route discovery phase, source transmits the route request message including four corners of the requested zone and an intermediate node decides whether to transmit the message or not. Using the position from scheme LAR, **Scheme 2** uses distance to define the requested zone. The intermediate node transmits if it is adjacent to the destination's previous location than the node transmitting the request packet; it is repeated until it is not received by its intended recipient [4] [44].

2.3.3. Geography Source Routing (GSR)

In GSR, source node computes the shortest path to the destination using dijkstra's algorithm based on distance metrics. It computes the distance from the source to intermediate nodes through which data is to be forwarded [45]. The source node queries for the location and floods the packet to the nodes, which wastes bandwidth. Spatially Aware Routing: It uses the GSR packet forwarding strategy to overcome the problem of recovery strategy in GPSR. It calculates the shortest path using dijkstra's algorithm. Source sets GSR consists of a list of intermediate nodes embedded in the header of all data packets by a source. Each forwarding node maps the position of its neighbors into graph nodes and chooses the next node having the shortest path from the destination, and then the packet will be forwarded to the next hop, which moves the data closer to the destination [4].

3. Analysis of MANET Routing Protocols

In this section, an analysis of the reviewed MANET routing protocols in terms of their characteristics, operation, strengths and weaknesses is presented. This section also highlights the drawbacks of these routing protocols to identify the areas that can be improved

3.1. Comparisons between MANET Routing Protocols

This subsection presents the comparison between the routing protocols reviewed in Section 2 above. **Table 1** presents a general comparison between the three categories of MANET routing protocols, while **Tables 2-4** present the comparison between the proactive, reactive and geographical MANET routing protocols respectively. **Table 1** summarizes the comparison between the Proactive, Reactive and Geographical MANET routing protocols discussed in this section. The comparison is done with respect to routing structure, availability of route, traffic control volume, periodic updates, control overhead, route acquisition delay, storage requirements, bandwidth requirement, power requirement, scalability problem, handling effects of mobility and quality of service support.

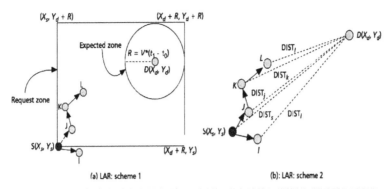

Figure 4. *LAR:* limited flooding of route request: a) scheme 1: expected zone; b) scheme 2: closer distances [43].

Table 1. Comparison between proactive, reactive and geographical MANET routing protocols.

Routing class	Reactive [46]	Geographical	Proactive [46]
Routing structure	Mostly flat, except cluster-based routing	Greedy forwarding routing	Both Flat and hierarchical structures
Availability of route	Determined when needed	Always available	Always available
Control Traffic volume	Lower than proactive routing protocols	Generate less control traffic	Usually high
Periodic updates	Not required. Some nodes may require periodic beacons.	Periodic beacons	Yes, some may use conditional.
Route acquisition delay	High	Low	Low
Storage Requirements	Depends on the number of routes kept or required. Usually lower than proactive protocols	The storage will be high since each node stores the locations	High
Bandwidth requirement	Low	High	High
Power requirement	Low	Low	High
Scalability	Source routing protocols up to few hundred nodes. Point-to-point may scale higher.	Limited Scalability problem	Usually up to 100 nodes.
Handling effects of mobility	Usually updates Associativity-Based Routing introduced localised broadcast query. AODV uses local route discovery	Constantly changing	Occur at fixed intervals and alters periodic updates based on mobility
Quality of service support	Few can support QoS , Although most support shortest path	Provide a node location service	Mainly shortest path as the QoS metric
Weaknesses	Have high latency, Flooding can lead to network clogging.	Short life of nodes in the networks due to the frequency of communication in each node.	Unsuitable for reconfigurable wireless ad-hoc network environment and not suitable for large networks.
Strengths	Reduce the overheads because it does not need to maintain up-to-date information about the network.	Suitable for sensor networks. The mobility support can be facilitated.	Control traffic are constant, and routes are always available.

Table 2 summarizes the comparison between the three proactive routing protocols. The comparison is done with respect to Multicast, number of routing tables used, and the frequency of updates

Table 3 summarizes the comparison between the three reactive routing protocols. The comparison is done with respect to multiple routes, route metric method, and the route reconfiguration strategy.

Table 4 summarizes the comparison between the three geographical routing. The comparison is done with respect to routing structure, number of routing tables used, and the frequency of updates.

Table 2. Comparison between proactive routing protocols [46].

Routing class	OLSR	DSDV	WRP
Multicast	No	Yes	No
Number of tables	five	Two	Four
Frequency of updates	Periodic	Periodic and as required	Periodic
Weaknesses	For the control message increases when the numbers of mobile nodes are increased, need higher processing power, 2-hop neighbor knowledge required	Unsuitable for highly dynamic networks, requires battery power, High overhead	Consumes a lot of bandwidth, as well as the power of each node is required to stay life all times, High Memory Overhead
Strengths	minimizes the size of protocol message and the number of rebroadcasting nodes during each route, Reduced control overhead and connection	Small amounts of bandwidth, loop free	provides the faster route convergence, Loop free

Table 3. Comparison between of reactive routing protocols [29].

Routing class	AODV	TORA	ABR
Multiple routes	No	Yes	No
Route metric method	Freshest and Shortest path	Shortest path or next available	Strongest Associatively and Shortest path
Route reconfiguration strategy	Erase route then source notification or local route repair	Link reversal and Route Repair	Localized Broadcast Query
Weaknesses	Periodic beaconing result in excessive bandwidth consumption, the intermediate nodes might result in inconsistent routes, Scalability problems, Large delays and Hello messages	In large networks the overhead, consume a large bandwidth, Temporary routing loops and Overall complexity	Lack loops, deadlock, and packet duplicates, Scalability problems, High Overhead and Overall complexity
Strengths	AODV has potentially less routing overheads, Adaptive to highly, Dynamic topologies and Low overhead	Able to rapidly build routes and decrease the communication's overhead, Multiple routes	Longer-lived routes, Route stability

Table 4. Comparison between Geographical routing protocols [10] [29].

Routing class	GSR	GPSR	LAR
Routing structure	Flat	Periodic beaconing	Location-based
Routing metric	Shortest Path	Closest distance	Shortest Path
Communication overhead	High	High	lower
Weaknesses	High delay at high mobility and High Memory Overhead	Delay increases at high mobility, generates a large number of control packets for high speeds, don't have better performance as the inter-beacon interval and group Leader is Single Point of Failure and low packet delivery ratio	Request the destination location, which might create contacting or tracking problem and Control complexity is higher than GPSR.
Strengths	Shortest path to the destination and Localized updates	Guarantees a good Packet Delivery Ratio especially in the high density of nodes, Keeps a good rate of delivery in networks with high mobility, generates routing protocol traffic in quantity independent of the length of the routes through the network and low data forwarding overhead and local maxima can be found easily	Decrease the search space results in little route discovery messages and it has minimized the size of the route discovery process by defining the range of the destination node.

3.2. Drawbacks of MANET Routing Protocols

Many researches have compared and analysed the characteristics and functionality of MANET routing protocols within these three categories. They observe that proactive routing protocols are unsuitable for reconfigurable wireless ad-hoc network environment due to the excessive use of the network capacity to maintain an up-to-data topological map on the entire network during the movement of nodes and network topology changes. In OLSR, the overhead for the control message increases when the numbers of mobile nodes increase. It will also need higher processing power compared to other protocols when trying to look for other routes [47]. OLSR [48]-[50], depends on the synchronized clocks among the nodes in the ad-hoc network. The reliance of this protocol upon the intermediate lower layers for selected functionalities assumes that the connected status sensing and neighbor discovery that are needed for packet delivery and address resolution are all readily available.

WRP updates the message transmission to its adjacent neighbors, and the nodes within the response list of the updated message will acknowledge its receipt to the class of path algorithm. WRP provides the faster route convergence [51]. DSDV [27] [52], requires constant updating to its routing tables which needs battery power and small amount of bandwidth, despite the network being idle. Upon alteration of the topology of the network, a unique sequence number is needed prior to network converging which renders DSDV unsuitable for highly dynamic networks.

Reactive routing protocols have high latency due to the need to look for a route to the destination before data can be sent. Flooding can lead to network clogging, while RREP, RREQ & RERR messages lead to control overheads [53]-[55]. TORA is able to rapidly build routes and decrease the communication overheads via the localization of the response to topological changes [30], which results in decreased processing and bandwidth usage of TORA protocol in terms of route creation, route maintenance and route erasure. In AODV, the intermediate nodes might result in inconsistent routes if the source sequence number is very old and the intermediate nodes have a higher, but not the latest destination sequence number, thereby having stale entries. Furthermore, the generation of many RREP packets in response to a single RREQ packet might lead to heavy control overheads. Another limitation of AODV is the fact that periodic beaconing result in excessive bandwidth consumption [56] [57].

Geographical routing is suitable for sensor networks, where data aggregation is utilized to minimize transmissions to base station via the elimination of redundancy between packets from multiple sources. GPSR has low overhead for data forwarding and local maximums are easily found. The GSR and the source node compute the shortest path to the destination using Dijkstra's algorithm, based on distance metrics [38]. It computes distance from the source to intermediate nodes through which data is to be forwarded. In LAR, when the source node intends to dispatch a data packet, it will first request for the location of the destination which might create contacting or tracking problem [44].

4. Conclusion

This paper presents a comparative study of routing protocols in mobile ad-hoc networks. These protocols are divided into three: proactive or table-driven, reactive or on-demand, and geographical routing protocols. For each of these classes, we have reviewed several representative protocols. Each routing protocol has unique features. The main factor that distinguishes the protocols is the method of determining routes within source destination pairs. The drawbacks, strengths and weaknesses of each protocol have also been examined in this paper. Reactive routing protocols suffer from longer delays and proactive routing protocols have higher overhead. The geographical routing protocol is very suitable for sensor networks whereby data aggregation is effective in minimizing transmission towards the base station via the elimination of redundancy among packets of multiple sources.

References

[1] Singh, G. and Singh, A. (2012) Performance Evaluation of Aodv and Dsr Routing Protocols for Vbr Traffic for 150 Nodes in MANETs. *International Journal of Computational Engineering Research* (ijceronline. com), **2**, 1583-1587.

[2] Kaur, R. and Rai, M.K. (2012) A Novel Review on Routing Protocols in MANETs. *Undergraduate Academic Research Journal* (*UARJ*), **1**, 103-108.

[3] Dhenakaran, S.S. and Parvathavarthini, A. (2013) An Overview of Routing Protocols in Mobile Ad-Hoc Network. *International Journal of Advanced Research in Computer Science and Software Engineering*, **3**, 251-259.

[4]   Malhotra, S. and Gill, N.S. (2014) Analysing Geographic Based Routing Protocols in MANETs. *International Journal of Computer Science and Mobile Computing*, **3**, 1068-1073

[5]   Panda, I. (2012) A Survey on Routing Protocols of MANETs by Using Qos Metrics. *International Journal of Advanced Research in Computer Science and Software Engineering*, **2**, 120-129.

[6]   Abolhasan, M., Wysocki, T. and Dutkiewicz, E. (2004) A Review of Routing Protocols for Mobile Ad Hoc Networks. *Ad Hoc Networks*, **2**, 1-22. http://dx.doi.org/10.1016/S1570-8705(03)00043-X

[7]   Mbarushimana, C. and Shahrabi, A. (2007) Comparative Study of Reactive and Proactive Routing Protocols Performance in Mobile Ad Hoc Networks. *21st International Conference on Advanced Information Networking and Applications Workshops, AINAW'07*, Niagara Falls, 21-23 May 2007, 679-684.

[8]   BR, A.K., Reddy, L.C. and Hiremath, P.S. (2008) Performance Comparison of Wireless Mobile Ad-Hoc Network Routing Protocols. *IJCSNS International Journal of Computer Science and Network Security*, **8**, 337-343.

[9]   Guo, J. and Wang, A. (2014) Study on Integration OLSR Protocol in Mobile Ad Hoc Network. *Proceedings of the 9th International Symposium on Linear Drives for Industry Applications*, **4**, 701-708.

[10]   Bali, S., Steuer, J. and Jobmann, K. (2008) Capacity of Ad Hoc Networks with Line Topology Based on UWB and WLAN Technologies. *Wireless Telecommunications Symposium*, Pomona, 24-26 April 2008, 17-24. http://dx.doi.org/10.1109/WTS.2008.4547538

[11]   Lol, W.G. (2008) An Investigation of the Impact of Routing Protocols on MANETs Using Simulation Modelling. Auckland University of Technology, Auckland.

[12]   Perkins, C.E. and Bhagwat, P. (1994) Highly Dynamic Destination-Sequenced Distance-Vector Routing (DSDV) for Mobile Computers. *ACM SIGCOMM Computer Communication Review*, **24**, 234-244.

[13]   Boukerche, A., Turgut, B., Aydin, N., Ahmad, M.Z., Bölöni, L. and Turgut, D. (2011) Routing Protocols in Ad Hoc Networks: A Survey. *Computer Networks*, **55**, 3032-3080. http://dx.doi.org/10.1016/j.comnet.2011.05.010

[14]   Bakht, H. (2011) Survey of Routing Protocols for Mobile Ad-Hoc Network. *International Journal of Information and Communication Technology Research*, **1**, 258-207.

[15]   Royer, E.M. and Toh, C.K. (1999) A Review of Current Routing Protocols for Ad Hoc Mobile Wireless Networks. *IEEE Personal Communications*, **6**, 46-55. http://dx.doi.org/10.1109/98.760423

[16]   Wan, T., Kranakis, E. and Van Oorschot, P.C. (2004) Securing the Destination-Sequenced Distance Vector Routing Protocol (S-DSDV). In: *Information and Communications Security*, Springer, Berlin, 358-374.

[17]   Dhenakaran, D.S. and Parvathavarthini, A. (2013) An Overview of Routing Protocols in Mobile Ad-Hoc Network. *International Journal of Advanced Research in Computer Science and Software Engineering*, **3**, 251-259.

[18]   Humblet, P.A. (1991) Another Adaptive Distributed Shortest Path Algorithm. *IEEE Transactions on Communications*, **39**, 995-1003. http://dx.doi.org/10.1109/26.87189

[19]   Rajagopalan, B. and Faiman, M. (1991) A Responsive Distributed Shortest-Path Routing Algorithm within Autonomous Systems. *Journal of Internetworking: Research and Experience*, **2**, 51-69.

[20]   Chowdhury, S. A., Uddin, M. A., and Al Noor, S. (2012) A Survey on Routing Protocols and Simulation Analysis of WRP, DSR and AODV in Wireless Sensor Networks.

[21]   de Morais Cordeiro, C. and Agrawal, D.P. (2011) Ad Hoc and Sensor Networks: Theory and Applications. World Scientific, Singapore. http://dx.doi.org/10.1142/8066

[22]   Kumar, G.V., Reddyr, Y.V. and Nagendra, D.M. (2010) Current Research Work on Routing Protocols for MANET: A Literature Survey. *International Journal on Computer Science and Engineering*, **2**, 706-713.

[23]   Bhat, M.S., Shwetha, D. and Devaraju, J. (2011) A Performance Study of Proactive, Reactive and Hybrid Routing Protocols using Qualnet Simulator. *International Journal of Computer Applications*, **28**, 10-17.

[24]   Khatri, P., Rajput, M., Shastri, A. and Solanki, K. (2010) Performance Study of Ad-Hoc Reactive Routing Protocols. *Journal of Computer Science*, **6**, 1159-1163. http://dx.doi.org/10.3844/jcssp.2010.1159.1163

[25]   Mewada, S. and Kumar, U. (2011) Measurement Based Performance of Reactive and Proactive Routing Protocols in WMN. *International Journal of Advanced Research in Computer Science and Software Engineering*, **1**, 1-6

[26]   Toh, C.K. (1996) A Novel Distributed Routing Protocol to Support Ad-Hoc Mobile Computing. *Proceedings of the 1996 IEEE Fifteenth Annual International Phoenix Conference on Computers and Communications*, Scottsdale, 27-29 March 1996, 480-486.

[27]   Khiavi, M.V., Jamali, S. and Gudakahriz, S.J. (2012) Performance Comparison of AODV, DSDV, DSR and TORA Routing Protocols in MANETs. *International Research Journal of Applied and Basic Sciences*, **3**, 1429-1436.

[28]   Misra, P. (1999) Routing Protocols for Ad Hoc Mobile Wireless Networks. Courses Notes. http://suraj.lums.edu.pk/~cs678/papers/17_routing_for_ad_hoc.pdf

[29] Islam, M.S., Riaz, M.A. and Tarique, M. (2012) Performance Analysis of the Routing Protocols for Video Streaming over Mobile Ad Hoc Networks. *International Journal of Computer Networks & Communications* (*IJCNC*), **4**, 133-150. http://dx.doi.org/10.5121/ijcnc.2012.4310

[30] Pragati, E. and Nath, D.R. (2012) Performance Evaluation of AODV, LEACH & TORA Protocols through Simulation. *International Journal of Advanced Research in Computer Science and Software Engineering*, **2**, 85-89.

[31] Aujla, G.S. and Kang, S.S. (2013) Comprehensive Evaluation of AODV, DSR, GRP, OLSR And TORA Routing Protocols with Varying Number of Nodes and Traffic Applications over MANETs. *IOSR Journal of Computer Engineering*, **9**, 54-61. http://dx.doi.org/10.9790/0661-0935461

[32] Som, D.S. and Singh, D. (2012) Performance Analysis and Simulation of AODV, DSR and TORA Routing Protocols in MANETs. *International Journal of Recent Technology and Engineering* (*IJRTE*), **1**, 122-127.

[33] Chakeres, I.D. and Belding-Royer, E.M. (2004) AODV Routing Protocol Implementation Design. 24*th International Conference on Distributed Computing Systems Workshops*, 23-24 March 2004, 698-703. http://dx.doi.org/10.1109/ICDCSW.2004.1284108

[34] Das, S.R., Belding-Royer, E.M. and Perkins, C.E. (2003) Ad Hoc On-Demand Distance Vector (AODV) Routing. http://www.ietf.org/rfc/rfc3561

[35] Etorban, A.A. (2012) The Design and Performance Evaluation of a Proactive Multipath Routing Protocol for Mobile Ad Hoc Networks. Heriot-Watt University, Edinburgh.

[36] Perkins, C.E. and Royer, E.M. (1999) Ad-Hoc On-Demand Distance Vector Routing. *Second IEEE Workshop on Mobile Computing Systems and Applications*, *Proceedings*, *WMCSA*'99, New Orleans, 25-26 February 1999, 90-100.

[37] Karkazis, P., Leligou, H., Orphanoudakis, T. and Zahariadis, T. (2012) Geographical Routing in Wireless Sensor Networks. 2012 *International Conference on Telecommunications and Multimedia* (*TEMU*).

[38] Kumar, P., Chaturvedi, A. and Kulkarni, M. (2012) Geographical Location Based Hierarchical Routing Strategy for Wireless Sensor Networks. 2012 *International Conference on Devices, Circuits and Systems* (*ICDCS*), Coimbatore, 15-16 March 2012, 9-14.

[39] Sohraby, K., Minoli, D. and Znati, T. (2007) Wireless Sensor Networks: Technology, Protocols, and Applications. John Wiley & Sons, Hoboken. http://dx.doi.org/10.1002/047011276X

[40] Battula, R.S. and Khanna, O. (2013) Geographic Routing Protocols for Wireless Sensor Networks: A Review. *International Journal of Engineering and Innovative Technology* (*IJEIT*), **2**, 39-42.

[41] Karp, B. and Kung, H.T. (2000) GPSR: Greedy Perimeter Stateless Routing for Wireless Networks. *Proceedings of the 6th Annual International Conference on Mobile Computing and Networking*, Boston, 6-11 August 2000, 243-254. http://dx.doi.org/10.1145/345910.345953

[42] Ko, Y.B. and Vaidya, N.H. (2000) Location-Aided Routing (LAR) in Mobile Ad Hoc Networks. *Wireless Networks*, **6**, 307-321. http://dx.doi.org/10.1023/A:1019106118419

[43] Mikki, M.A. (2009) Energy Efficient Location Aided Routing Protocol for Wireless MANETs. arXiv preprint arXiv:0909.0093.

[44] Hong, X., Xu, K. and Gerla, M. (2002) Scalable Routing Protocols for Mobile Ad Hoc Networks. *IEEE Network*, **16**, 11-21. http://dx.doi.org/10.1109/MNET.2002.1020231

[45] Liu, L., Wang, Z. and Jehng, W.K. (2008) A Geographic Source Routing Protocol for Traffic Sensing in Urban Environment. *IEEE International Conference on Automation Science and Engineering*, *CASE* 2008, Arlington, 23-26 August 2008, 347-352.

[46] Mohseni, S., Hassan, R., Patel, A. and Razali, R. (2010) Comparative Review Study of Reactive and Proactive Routing Protocols in MANETs. 4*th IEEE International Conference on Digital Ecosystems and Technologies* (*DEST*), Dubai, 13-16 April 2010, 304-309. http://dx.doi.org/10.1109/DEST.2010.5610631

[47] Hassnawi, L., Ahmad, R., Yahya, A., Aljunid, S. and Elshaikh, M. (2012) Performance Analysis of Various Routing Protocols for Motorway Surveillance System Cameras' Network. *International Journal of Computer Science Issues*, **9**, 7-21.

[48] Clausen, T., Jacquet, P., Adjih, C., Laouiti, A., Minet, P., Muhlethaler, P., Viennot, L., *et al.* (2003) Optimized Link State Routing Protocol (OLSR). http://dx.doi.org/10.17487/rfc3626

[49] Johnson, D.B. (1994) Routing in Ad Hoc Networks of Mobile Hosts. *First Workshop on Mobile Computing Systems and Applications*, *WMCSA* 1994, Santa Cruz, 8-9 December 1994, 158-163.

[50] Reddy, P.N., Vishnuvardhan, C. and Ramesh, V. (2013) An Overview on Reactive Protocols for Mobile Ad-Hoc Networks. *International Journal of Computer Science and Mobile Computing*, **2**, 368-375.

[51] Srivastava, S., Yadav, A. and Kumar, A. (2013) Impact of Node Mobility of Routing Protocols on MANET. *International Journal of Scientific & Engineering Research*, **4**, 1424-1431.

[52] Janakiraman, G., Raj, T.N. and Suresh, R. (2013) AODV, DSDV, DSR Performance Analysis with TCP Reno, TCP New Reno, TCP Vegas on Mobile Ad-Hoc Networks Using NS2. *International Journal of Computer Applications*, **72**, 1-7.

[53] Gupta, A.K., Sadawarti, H. and Verma, A.K. (2011) A Review of Routing Protocols for Mobile Ad Hoc Networks. *SEAS Transactions on Communications*, **10**, 331-340.

[54] Gupta, A.K., Sadawarti, H. and Verma, A.K. (2011) Review of Various Routing Protocols for MANETs. *International Journal of Information and Electronics Engineering*, **1**, 251-259. http://dx.doi.org/10.7763/IJIEE.2011.V1.40

[55] Gupta, N. and Shrivastava, M. (2013) An Evaluation of MANET Routing Protocol. *International Journal of Advanced Computer Research*, **3**, 165-170.

[56] Gupta, A. and Pradhan, M. (2013) A Comparative Study of Current Routing Protocol in Wireless Ad-Hoc Network. *International Journal of Engineering Research and Technology*, **2**, 144-148.

[57] Malhotra, R. and Kaur, S. (2011) Comparative Analysis of AODV and DSR Protocols for Mobile Adhoc Networks. *International Journal of Computer Science & Engineering Technology*, **1**, 330-335.

10

A Defense Framework against DDoS in a Multipath Network Environment

Ahmed Redha Mahlous

Department of Computer and Information Sciences, Prince Sultan University, Riyadh, Saudi Arabia
Email: armahlous@psu.edu.sa

Abstract

The Internet is facing a major threat, consisting of a disruption to services caused by distributed denial-of-service (DDoS) attacks. This kind of attacks continues to evolve over the past two decades and they are well known to significantly affect companies and businesses. DDoS is a popular choice among attackers community. Such attack can easily exhaust the computing and communication resources of its victim within a short period of time. Many approaches to countering DDoS attacks have been proposed, but few have addressed the use of multipath. In this paper, we analyze, how multipath routing based solutions could be used to address the DDoS problem. The proposed framework traces back the attack to its source and blocks it. It also calculates multiple paths to the attacker (if they exist) and alerts all gateways near the attacker to block possible traffic originating from this source in case another path(s) is (are) later used to attack the victim again. We demonstrate that our scheme performs better that other single path schemes.

Keywords

DDoS, Multipath, Filtering, Traceback

1. Introduction

A distributed denial-of-service (DDoS) attack is a serious threat to the security of the Internet. It consists of exhausting the bandwidth, processing capacity or memory of a targeted machine or network. It is a distributed, cooperative and large-scale attack. It has been widespread on wired [1] and wireless networks [2].

Attackers usually hide their identity and DDoS is an example of anonymous attacks where currently there is no obvious way to prevent or trace them. Practically it is impossible to prevent all DDoS attack(s) that originate from Internet; however we can at least find a mechanism to identify the source(s) of the attack in situation where prevention fails. In this context security community proposes IP traceback techniques.

In this paper we present a filtering framework mechanism for a DDoS attack, which is similar to that one used in [3]. However, our new framework computes multiple paths, if they exist, to the attack source(s), then it generates an alert to the nearest gateway(s) of those paths, to block all traffic originating from the attack source(s), thing that has not been investigated in [3]. In this way our proposed framework prevents the attacker from using other available paths to attack the same destination.

Our source address spoofing solution is a hardware-friendly variant of the IP route record (RR) technique. It uses recording packet route information to block attack traffic from different paths. The marking techniques used are different from traditional ones [4] [5] (which do not provide an explicit recorded route in each packet). The aim of our multipath traffic filtering (MTF) system is to use the recorded route on each incoming packet to block that traffic at the last point of trust on each attack path, which means blocking untrusted traffic at the nearest possible point from the attack source(s).

The remainder of the paper is organized as follows: Section 2 presents related work, while Section 3 describes the MTF components. Section 4 describes the MTF protocol in detail. MTF simulation and performance are evaluated in Section 5. And Section 6 concludes the paper.

2. Related Works

Many techniques against DDOS have been studied in recent years, which lead to the proposal of a defense mechanism to be implemented on routers [6]-[10], servers [11] [12], or both servers and clients [13]-[16].

Authors in [17] proposed a targeted filtering mechanism that blocks stateful DDoS attack called targeted filtering. Filters are established at a firewall and automatically converges the filters to the flooding sources [18]; authors studied the impact of application layer Denial-of-Service (DoS) attacks that targets web services. They proposed an adaptive system that detect against application layer attacks such as XML and HTTP. Furthermore, authors state that the system is also capable of detecting malicious request, spoofing and regular flooding attacks. It is intended to be deployed in a cloud environment where it can transparently protect the cloud broker and cloud providers.

A Bandwidth DDoS (BW-DDoS) attack was the study of [19]. Authors pointed that, this type of attack can disrupt network infrastructure operation by causing congestion, this is due to the increase of total amount traffic inducing connectivity degradation or more than that, loss of connection between the Internet and victim networks or even whole autonomous systems (ASs). Authors presented some type of attack agent, attack mechanism, protocol manipulation, and attack target and response mechanism.

A New approach to IP traceback named Deterministic Flow Marking (DFM) was presented in [20]. DFM allows the victim to traceback the origin of spoofed source addresses up to the attacker node even if this later is hidden behind a NAT or a proxy server. It has the capability to trace simultaneous distributed attacks in near real time while providing authentication according to authors.

Authors in [21] argued that the existing application layer methodologies couldn't detect DDoS attack. They proposed a set of algorithms that are capable of detecting and blocking DDoS attacks whilst allowing through legitimate user traffic, including flash traffic. Discriminating the attack flows from the flash crowds was also the study of [22] and [23].

To the best of our knowledge all traceback techniques that were studied confines to only one attack path. In this paper we propose a novel multipath traceback technique to defend against DDoS attack.

3. MTF Components

3.1. Path Recording

MTF appends the node's address to the end of the packet as it travels through the network from the attacker to victim. Consequently, every packet received by the victim arrives with a complete ordered list of the routers that it traverses providing a built-in attack path. So a path fingerprint is embedded in each packet, enabling a victim to identify packets traversing the same paths through the Internet on a per packet basis, regardless of source IP address spoofing. In this way the victim needs only classify a single packet as malicious to be able to trigger a filtering action that will be applied to subsequent packets with the same marking (all packets traversing the same path carry the same marking). This approach is very lightweight, both on the routers for marking, and on the victims for decoding and filtering.

We assume that each router that participates in our scheme has a capability to add its IP address to the packets that traverse it and is MTF capable. We also assume that only the border router participates in this scheme and not the core router connected to the Internet backbones. As a result, each packet carries the identities of a sub-list of the border routers that forwarded it.

To avoid the performance overhead of a traditional IP route record, the path header (PH) is introduced at the beginning of the IP payload and it works as a "shim" protocol between the IP and transport layers.

The PH used follows the recommendations stated in [24] and is composed of three fields: the path is a list of (initially empty) slots; the size is the total number of slots in the path; the pointer points to the first empty slot in the path. The size of the number of slots represents the maximum distance that a packet can reach.

As the length of route cannot be predicted and each hop increases the packet header size (that grows linearly with hop count) which can lead to fragmentation due to the limit of the maximum transmission unit (MTU), we set the maximum hop to 15 as used by the well-known RIP network protocol [25]. If a packet reaches a router beyond this number it is dropped (the router does not have an empty slot to write its IP address).

When a packet traverses an MTF-enabled router it does the following: (1) inserts a PH header in the packet; (2) writes its own IP address and random value in the first slot; and (3) sets the pointer to point to the second slot. Each subsequent MTF-enabled border router that ingresses the packet into a new AS: (1) reads the pointer; (2) writes its own IP address and random value in the indicated slot; and (3) increments the pointer to point to the next slot. If no room is left in the PH header (after 16 hops), the router drops the packet.

3.2. Multi-Path Calculation

The victim's gateway will calculate multiple shortest paths to the source of attack (if they exist) and locate the nearest border gateway(s) to the attacker. To calculate the multiple shortest paths (if they exist), we use a modified variant of Dijkstra's algorithm [26] [27], as it is the most widely used algorithm in present day routers. By choosing the shortest paths we aim to send a warning message (blocking request) as fast as possible.

3.3. Flow Classification

When a packet crosses an MTF-enabled router, its recorded route includes an authentic (non-spoofed) suffix. Specifically, the last "n" components of the recorded route are authentic, when the last n border routers crossed by the packet are MTF-enabled, and there is no malicious node on the path that interconnects them using malicious node detection scheme [28].

The recorded path of a packet is defined as a sequence of IP addresses that corresponds to: the packet's source, the list of border routers included in the PH header, and the packet's destination. Packets are marked as belonging to the same flow if they share a common recorded path suffix. This procedure enables a receiver to identify distinct incoming traffic flows in the face of source address spoofing.

A flow F with recorded path P is denoted as FP. A DDoS node victim uses a filtering policy to classify incoming traffic in different flows, and then decides which one to apply a filtering procedure to and either reject it or accept it. The operation of the policy module depends on the specific service run by the victim and is outside the scope of this paper. The blocking of any undesired flow that may come from different paths and enforcing such a policy is achieved by deploying our proposed mechanism (MTF).

It is up to the policy module to classify incoming traffic in multiple flow levels. For example, consider that in **Figure 1** an attack source, PC 1, is sending high-rate traffic to the victim, SERVER. If network Net1 prevents source address spoofing, SERVER can easily identify F1 [PC1, R1 R3, SERVER] as a high-rate flow, and thus undesired. If PC1 is able to spoof multiple source IP addresses, SERVER can only identify F2 [*, R1, R3, SERVER] as the undesired flow.

Once the policy module identifies an undesired flow, it sends a filtering request to the local MTF process.

4. Basic MTF Protocol

4.1. Terminology

- The recorded path P of an undesired flow (**Figure 2**) has the form [PC1, R1, ...R3, SERVER], where PC1 is the attack source, *i.e.* the node thought to be generating the undesired traffic; if PC1 = *, all traffic through R1 is undesired.

- R1 is the attack gateway, *i.e.* the border router thought to be closest to PC1.
- R3 is the victim's gateway, *i.e.* the border router closest to the victim.
- SERVER is the victim.

We assume that the only node affected by the attack is SERVER; e.g. if this is a flooding attack, the only part of the network that is congested is the tail-circuit from R3 to SERVER. If R3 were also affected, it itself would be the victim, and its closest upstream border router would be the victim's gateway. As the majority of DDoS attacks are performed using TCP as summarized in **Table 1** [29], we assume that only TCP packets are used in our model.

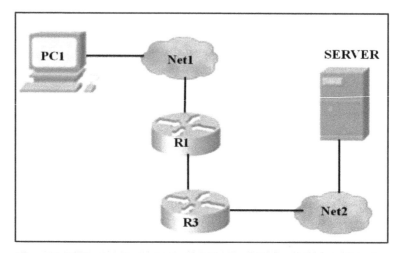

Figure 1. MTF network.

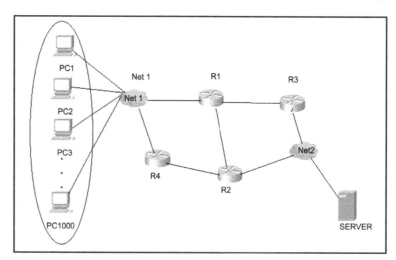

Figure 2. DDoS network attack scenario.

Table 1. Distribution of DDoS attack types.

Protocols	Ratios
TCP	90% - 94%
UDP	2.4% - 5%
ICMP	2.1% - 2.6%
Other	2.06% - 2.93%

4.2. Blocking Close to the Attack Source

MTF involves 4 units:

1) When an undesired flow(s) arrives at the victim SERVER, it sends a filtering request to R3, specifying an undesired flow F.

2) The victim's gateway R3:

(a) Installs a temporary filter to block F for Ttmp seconds.

(b) Initiates a 3-way handshake with R1 (**Figure 3**).

(c) Removes its temporary filter, upon completion of the handshake.

3) The attack gateway R1:

(a) Responds to the 3-way handshake.

(b) Installs a temporary filter to block F for Ttmp seconds, upon completion of the handshake.

(c) Sends a filtering request to the attackers' source (PC1…PC1000), to stop F for Tlong >> Ttmp minutes.

(d) Removes its temporary filters if the attackers comply within Ttmp seconds; otherwise, it disconnects the attackers.

4) the attack source stops F for Tlong minutes or risks disconnection.

When R1 blocks traffic it sends a positive message that the traffic has been blocked. During this time, R3 calculates whether there are other paths to the attacker. If yes, it sends a message to the gateways(s) closest to the attacker (R4 in **Figure 2**) warning it of a possible attack from a source attack. The gateway(s) starts monitoring traffic and blocking any packets coming from the warned sources. R4 will install a filter to block any future traffic from the attack sources for time T_{block}. T_{block} will have the same value as T_{temp}.

If the victim's warning message arrives after the attacker has used another path, some unwanted traffic might arrive at the victim, but progressively they are blocked after the victim carries out the same process described previously to block the attack.

If the warning message arrives before the attackers use another path(s), the victim, in this manner, has blocked probable attacks. Thus the attackers will have less chance to attack SERVER using other paths after being blocked.

When SERVER receives the attack, it sends a message to R1 with a suggested rate limit value. Based on the requirements in the message, R1's system will decrease the rate limit value exponentially. After the traffic at the victim's end has returned to normal for a period, an update message is sent to the source end asking it to increase the rate limit value linearly.

If the defense system has not found any anomalous changes at the victim's end since the update message, a cancel message to remove the rate limit at the source end is sent.

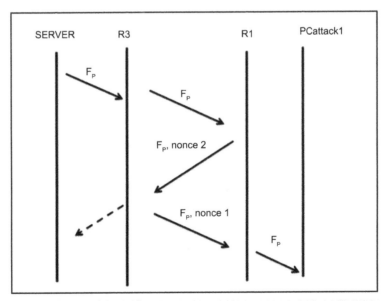

Figure 3. Three-way handshake.

However, if the volume of attack traffic still increases aggressively, an update message will be sent again to block the traffic coming from the attacker.

4.3. Preventing Man-in-the-Middle Attack

In order to secure communication between nodes and trigger a filter in the gateway near the source of the attack we use a 3-way handshake as shown in **Figure 3**.

The 3-way handshake is depicted in **Figure 3**: in which R3 sends a request to R1 to block F; R1 sends SERVER a message that includes F and a nonce; R3 intercepts the message and sends it back to R1.

R3 proves its location on the path to SERVER by intercepting the nonce sent to SERVER. This prevents malicious node H, located off the path from R1 to R3 (**Figure 4**), from causing a filter to be installed at R1 and block traffic to SERVER. By picking a sufficiently large and properly random value for the nonce, it can be made arbitrarily difficult for H to guess it. To defend against forced delay attack, the nonces are tied in with short response time outs.

To avoid a buffering state on incomplete 3-way handshakes, R1computes the nonce as follows:

$$\text{Nonce1} = \text{Hk}(F) \tag{1}$$

where Hk is a keyed hash function.

To verify the authenticity of a completion message, R1 just hashes the flow included in the message and compares the result to the nonce included in the message.

For example, in **Figure 4**, traffic sent by Net1 to Net2 has recorded path [R1:RND1, R3:RND2, Net2], where RND1 and RND2 are random values inserted by R1 and R3 respectively. To avoid keeping per destination states, each router computes the random value to insert in each packet as follows:

$$R = \text{Hk}(D) \tag{2}$$

where D is the packet's destination.

4.4. Filter Timeout Values

The goal of a temporary filter on the victim's gateway is to block an undesired flow until the corresponding handshake is complete. Considering that Internet roundtrip times range from 50 to 200 msec [30], we take the approximate average of $T_{tmp} = 1$ sec as a reasonable choice since it offers a safe value for a round trip time.

The choice of the long-term filter timeout T_{long} involves the following trade-off: An attack source PC1 is typically an innocent end-host compromised through a worm. A large T_{long} of, say, 30 minutes guarantees that the victim SERVER will not receive any undesired traffic from the corresponding attack PC1 for at least 30

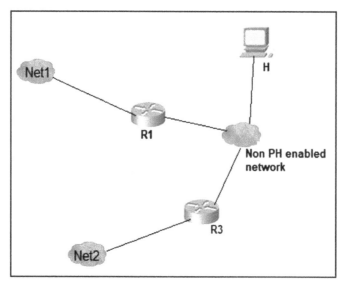

Figure 4. Man in the middle.

minutes; on the other hand it also guarantees that SERVER will not receive any traffic at all from PC1, even if PC1 is appropriately patched before T_{long} expires.

5. Simulation and Performance Results

5.1. Filtering Gain

In this section we use a NS-2 simulator [3] to conduct a set of simulations to evaluate the performance of the proposed MTF framework in the presence of DDoS attacks with different scenarios.

Figure 5 shows the experimental topology. Trusted users (PC 1 to PC 100) generate a TCP based FTP flow with a packet size of 1000 bytes. While DDoS traffic is generated by PCattack (PCattack 1 to PCattack 1000) by sending packets of 50 bytes in size. The victim's gateway uses up to 10,000 filters to protect the victim. For each scenario we plot the bandwidth of the attack traffic that reaches the victim as well as the victim's goodput as a function of time, *i.e.* we show how fast attack traffic is blocked and how much of the victim's goodput is restored. In all scenarios T_{temp} = 1 sec and T_{long} = 2 min. The choice for such value (T_{long} = 2 min) is to reduce the event handling overhead for the simulation run.

5.1.1. Scenario 1: Attackers Use One Path Only

The victim receives a flooding attack from 1000 attack sources. The bandwidth of the attack (before defense) is 1 Gbps. From Figure 6 we can see that at time t = 0 - 1 sec, before the attack, SERVER is receiving 80 Mbps of goodput, then when the attack starts from path 1 [R1-R3-SERVER] (Figure 5) the goodput is reduced to 12%

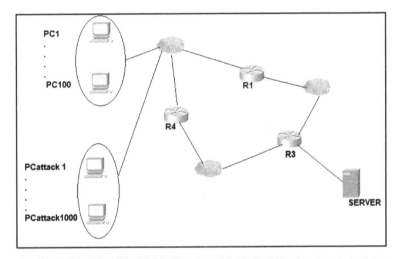

Figure 5. Experimental topology.

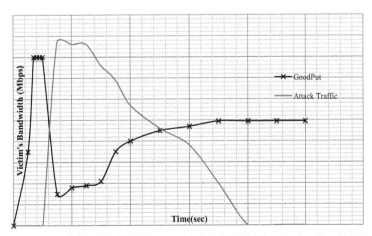

Figure 6. One path attack.

from its original value. At t = 2 sec, SERVER starts sending requests to the gateway to block the attack (it sends 100 filter requests); t = 2 - 8 half of goodput is restored progressively until all the attack traffic is blocked at t = 8.

5.1.2. Scenario 2: Attackers Use Multiple Path

In this scenario we assume that the victim receives a flooding attack from 1000 attack sources from another path [R4-R3-SERVER] (**Figure 5**). We can see clearly from **Figure 7** that at t = 0 - 1 SERVER receives normal traffic using 80 Mbps of goodput; at t = 1 - 2 the attack drives SERVER goodput to ~50% of its original; at t = 2 sec SERVER starts sending 100 filtering request/sec to the gateway, when this later blocks unwanted traffic, the attacker at t = 2 - 3 uses another path to flood the victim with unwanted traffic. The goodput decreases to 60% of its original, then at t = 3 sec SERVER's goodput is restored to ~78% of the original and the attacker is blocked.

R4 blocks 10 flows in 10 seconds and 1000 flows in 100 seconds. Without MTF, a router needs one thousand filters to block one thousands flows; these experiments demonstrate that, with MTF, R4 needs only one hundred filters to block one thousand flows.

In this experiment we demonstrated that MTF achieves a filtering response time equal to the one-way delay from the victim to its gateway.

Hence, MTF reduces the number of filters required to block a certain number of flows by two orders of magnitude. A critical improvement, since routers typically accommodate tens of thousands of filters, whereas DDoS attacks can easily consist of millions of flows.

5.2. Attack Path Reconstruction and False Positive Ratio

An important performance indicator for IP traceback is the number of packets that need to be collected for reconstructing an attack path. **Figure 9** shows the number of packets required to reconstruct attack paths. To better evaluate MTF's effectiveness; we repeat the same experiments with FMS [31] and AMS [32].

The result in **Figure 8** shows that the number of packets needed by MTF is the smallest among all PPM schemes. This is expected since MTF only needs to collect one marked packet along an attack path, whereas other schemes need multiple marked packets.

We are also concerned with the relationship between the false positive ratio and the number of attackers. **Figure 9** shows the false positive ratios versus the number of attackers.

The false positive ratio is the number of legitimate clients that are mistakenly recognized as attackers by the MTF divided by the number of legitimate clients. In each simulation we fix the total number of clients to 5000 and vary the number of randomly chosen attackers from 1 to 4101 in increments of 10. The results show that MTF incurs relatively high false positive ratios at the beginning of attack before gateways are involved in decreasing their rate traffic. The reason is that as soon as MTF detects an attack it will block all traffic from the attacker paths. This is because attacker and innocent packets reside behind the same gateway, and all packets coming from the latter are considered unwanted as soon as an attack occurs and the victim triggers the filtering process.

This cannot make things any worse, since most good traffic is being dropped anyway due to the flood; in

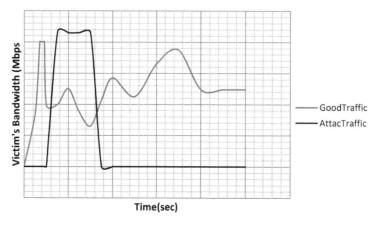

Figure 7. Multiple path attack.

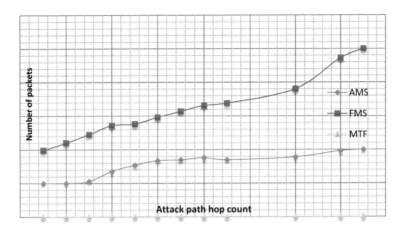

Figure 8. Numbers of packets.

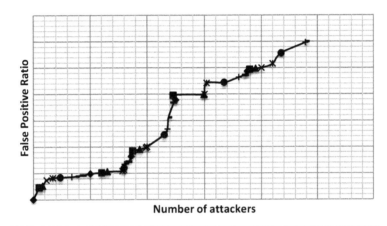

Figure 9. False positive ratio.

contrast it allows good traffic from other gateways to get through. Once the attacks are filtered, we notice that false positive ratio gradually decreases.

6. Conclusions

We presented a Multipath Traffic Filter (MTF), a mechanism for filtering highly distributed denial-of-service attacks. We showed that MTF can block a thousand undesired flows, while only requiring a hundred filters.

It also prevents abuse by malicious nodes seeking to disrupt the communications of other nodes. We showed that MTF has the capability of blocking unwanted flows from the same source from different paths, even before they can occur.

Our plans for future work include the development of a congestion control algorithm that will detect and control high bandwidth flow. Furthermore, we will look at mechanism for reducing computation and storage overhead induced by multipath calculation on control and data plane. Finally, we are interested in deploying such techniques on a larger scale network.

References

[1] Chonka, A., Xiang, Y., Zhou, W.L. and Bonti, A. (2010) Cloud Security Defense to Protect Cloud Computing against HTTP-DoS and XML-DoS Attacks. *Journal of Network and Computer Applications*, **34**, 1097-1107.

[2] Arunmozhi1, S.A. and Venkataramani, Y. (2011) DDos Attack and Defense Scheme in Wireless Ad Hoc Networks. *International Journal of Network Security & Its Applications (IJNSA)*, **3**, 182-187.

[3] Argyraki, K. and Cheriton, D.R. (2010) Active Internet Traffic Filtering: Real-Time Response to Denial-of-Service Attacks. *USENIX Annual Technical Conference*, Berkeley, CA, 10-10.

[4]   Yaar, A., Perrig, A. and Song, D. (2003) Pi: A Path Identification Mechanism to Defend against DDoS Attacks. *Proceedings IEEE Symposium on Security and Privacy Symposium*, 11-14 May 2003, 93-107.

[5]   Savage, S., Wetherall, D., Karlin, A. and Anderson, T. (2000) Practical Network Support for IP Traceback. *ACM SIGCOMM Computer Communication Review*, **30**, 295-306.

[6]   Ferguson, P. and Senie, D. (1998) Network Ingress Filtering: Defeating Denial of Service Attacks Which Employ IP Source Address Spoofing. IETF, Request For Comments 2267, United States.

[7]   Wang, H., Zhang, D. and Shin, K.G. (2002) SYN-Dog: Sniffing SYN Flooding Sources. *Proceedings of the 22nd International Conference on Distributed Computing Systems*, 421-428.

[8]   Park, K. and Lee, H. (2001) On the Effectiveness of Route-Based Packet Filtering for Distributed DoS Attack Prevention in Power-Law Internets. *Proceedings of the* 2001 *SIGCOMM Conference*, **31**, 15-26. http://dx.doi.org/10.1145/383059.383061

[9]   Sung, M. and Xu, J. (2003) IP Traceback-Based Intelligent Packet Filtering: A Novel Technique for Defending against Internet DDos Attacks. *IEEE Transactions on Parallel and Distributed Systems*, **14**, 861-872. http://dx.doi.org/10.1109/TPDS.2003.1233709

[10]   Mahajan, R., Bellovin, S.M., Floyd, S., Ioannidis, J., Paxson, V. and Shenke, S. (2002) Controlling High Bandwidth Aggregates in the Network. *ACM SIGCOMM Computer Communication Review*, **32**, 62-73. http://dx.doi.org/10.1145/571697.571724

[11]   Bernstein, D.J. (1997) SYN Cookies. http://cr.yp.to/syncookies.html

[12]   Xu, J. and Lee, W. (2003) Sustaining Availability of Web Services under Distributed Denial of Service Attacks. *IEEE Transactions on Computers*, **52**, 195-208. http://dx.doi.org/10.1109/TC.2003.1176986

[13]   Juel, A. and Brainard, J. (1999) Client Puzzles: A Cryptographic Countermeasure against Connection Depletion Attacks. *Proceedings of the Network and Distributed System Security Symposium*, *NDSS*'99, San Diego, February 1999.

[14]   Aura, T., Nikander, P. and Leiwo, J. (2000) Dos-Resistant Authentication with Client Puzzles. Cambridge Security Protocols Workshop 2000. Lecture Notes in Computer Science. Springer, Berlin, 200.

[15]   Dean, D. and Stubblefield, A. (2001) Using Client Puzzles to Protect TLS. *Proceedings of the Tenth USENIX Security Symposium*, Washington DC.

[16]   Wang, X. and Reiter, M.K. (2003) Defending against Denial-of-Service Attacks with Puzzle Auctions. *IEEE Symposium on Security and Privacy*, Los Alamitos, 11-14 May 2003, 78-92.

[17]   Chen, S.G., Tang, Y. and Du, W.L. (2007) Stateful DDos Attacks and Targeted Filtering. *Journal of Network and Computer Applications*, **30**, 823-840. http://dx.doi.org/10.1016/j.jnca.2005.07.007

[18]   Vissers, T., Somasundaram, T.S., Pieters, L., Govindarajan, K. and Hellinckx, P. (2014) DDos Defense System for Web Services in a Cloud Environment. *Future Generation Computer Systems*, **37**, 37-45. http://dx.doi.org/10.1016/j.future.2014.03.003

[19]   Geva, M., Herzberg, A. and Gev, Y. (2013) Bandwidth Distributed Denial of Service: Attacks and Defenses. *IEEE Security & Privacy*, **12**, 54-61. http://dx.doi.org/10.1109/MSP.2013.55

[20]   Foroushani, V.A. and Zincir-Heywood, A.N. (2013) Deterministic and Authenticated Flow Marking for IP Traceback. *IEEE 27th International Conference on Advanced Information Networking and Applications* (*AINA*), Barcelona, 25-28 March 2013, 397-404.

[21]   Sivabalan, S. and Radcliffe, P.J. (2013) A Novel Framework to Detect and Block DDos Attack at the Application Layer. *IEEE TENCON Spring Conference*, Sydney, 17-19 April 2013, 578-582. http://dx.doi.org/10.1109/TENCONSpring.2013.6584511

[22]   Sowkarthiga, P. and Suguna, N. (2013) Finding the DDoS Attacks in the Network Using Distance Based Routing. *International Conference on Current Trends in Engineering and Technology (ICCTET)*, *ICCTET*'13, Coimbatore, 3 July 2013, 410-412.

[23]   Tao, Y. and Yu, S. (2013) DDoS Attack Detection at Local Area Networks Using Information Theoretical Metrics. *TRUSTCOM*'13, *Proceedings of the* 2013 *12th IEEE International Conference on Trust, Security and Privacy in Computing and Communications*, Melbourne, 16-18 July 2013, 233-240. http://dx.doi.org/10.1109/TrustCom.2013.32

[24]   Anderson, T., Roscoe, T. and Wetherall, D. (2003) Preventing Internet Denial-of-Service with Capabilities. *Newsletter ACM SIGCOMM Computer Communication Review*, **34**, 39-44

[25]   Ioannidis, J. and Bellovin, S.M. (2002) Implementing Pushback: Router-Based Defense against DDos Attacks. *Proceedings of the Network and Distributed System Security Symposium*, San Diego, ISOC, Reston, VA.

[26]   Mahlous, R., Chaourar, B. and Fretwell, R.J. (2008) A Comparative Study Between Max Flow Multipath, Multi Shortest Paths And Single Shortest Path. In: *Proceedings of PGNet* 2008, PGNet, Liverpool.

[27] Walfish, M., Vutukuru, H., Karger, D. and Shenker, S. (2010) DDos Defense by Offense. *Journal of ACM Transactions on Computer Systems (TOCS)*, **28**.

[28] Lokanath, S. and Thayur, A. (2013) Implementation of AODV Protocol and Detection of Malicious Nodes in Manets. *International Journal of Science and Research (IJSR)*, **2**.

[29] Yaar, A., Perrig, A. and Song, D. (2004) SIFF: A Stateless Internet Flow Filter to Mitigate DDoS Flooding Attacks. *In Proceedings IEEE Symposium on Security and Privacy Symposium*, 9-12 May 2004, 130-143.

[30] Osipov, E., Kassler, A., Bohnert, T.M. and Masip-Bruin, X. (2010) Wired/Wireless Internet Communications. *8th International Conference on Wired/Wireless Internet Communications (WWIC)*, Luleå, 1-10 June 2010.

[31] Lee, W. and Xu, J. (2003) Sustaining Availability of Web Services under Distributed Denial of Service Attacks. *IEEE Transactions on Computers*, **52**, 195-208.

[32] Beitollahi, H. and Deconinck, G. (2011) A Cooperative Mechanism to Defense against Distributed Denial of Service Attacks. *IEEE 10th International Conference on Trust, Security and Privacy in Computing and Communications (TrustCom)*, Changsha, 16-18 November 2011, 11-20.

11

Dynamic Capacity Allocation in OTN Networks

Maria Catarina Taful[1], João Pires[1,2]

[1]Department of Electrical and Computer Engineering, Instituto Superior Técnico, University of Lisbon, Lisbon, Portugal
[2]Institute of Telecommunications, Instituto Superior Técnico, University of Lisbon, Lisbon, Portugal
Email: mariacatarinataful@gmail.com, jpires@lx.it.pt

Abstract

A dynamic Optical Transport Network (OTN) has the advantage of being able to adjust the connection capacity on demand in order to respond to variations on traffic patterns or to network failures. This feature has the potential to reduce operational costs and at the same time to optimize networks resources. Virtual Concatenation (VCAT) and Link Capacity Adjustment Scheme (LCAS) are two techniques that when properly combined can be used to provide improved dynamism in OTN networks. These techniques have been previously standardized in the context of Next Generation SDH/SONET networks. VCAT is used to tailor the capacity of network connections according to service requirements, while LCAS can adjust dynamically that capacity in a hitless manner. This paper presents an overview of the application of VCAT/LCAS techniques in the context of OTN. It explains in detail how these techniques can be employed to resize the connection capacity and analyses its use in network protection solutions. Furthermore, a detailed analysis of the time delays associated with different operations is provided and its application to some reference networks is undertaken. The obtained results provide an idea about the time delays of the capacity adjustment processes and define potential scenarios for implementing VCAT/LCAS techniques.

Keywords

OTN, VCAT, LCAS, Time Delay, Network Protection

1. Introduction

In order to support the constant growing of network traffic and the increasing heterogeneity of services/applications the transport infrastructure of telecommunication networks is facing a series of new challenges. The traffic growth imposes the usage of very-high bit rates (*i.e.* 40 and 100 Gb/s) and the rising variety of flows requests for more flexible, reliable and dynamic network designs. These networks must be capable, for example, to provide fast re-provisioning of services to accommodate traffic fluctuations and at the same time to respond as quickly as possible to network failures. The Optical Transport Network (OTN) technology is expected to be the right technology to handle these challenges [1] [2]. It has been standardized by ITU-T in Recommendation G. 709 [3], operates at layer 1 of the Open Systems Interconnection (OSI) communications model and it is itself subdivided into two layers: an electrical layer also called digital wrapper and an optical layer also called Dense Wavelength Division Multiplexing (DWDM) layer. The electrical layer is responsible for mapping client signals into entities called Optical Data Units (ODUs), as well as for multiplexing, switching and managing these entities, whereas the optical layer is responsible for generating, multiplexing, switching and managing optical channels. The ODUks (k = 0, 1, 2, 2e, 3, 4) are transport containers used to carry client signals between an end-to-end path and can be either of fixed or variable size. The containers of fixed size are standardized to support certain client signals. For example, ODU4 is intended to transport a 100 GbE signal. To obtain a container of variable size, there are two techniques available: Flexible Rate ODU (ODUflex) and Virtual Concatenation (VCAT) [4] [5].

In ODUflex, a certain number of Tributary Slots, each one with a granularity of approximately 1.25 Gb/s, are combined and the resulting structure is mapped into a fixed higher order ODUk to be transported as a single entity. On the other hand, VCAT is an inverse multiplexing technique, by which each payload container of a given traffic flow is segmented into smaller containers, which are logically combined to form a Virtual Concatenation Group (VCG) and transported independently of each other over the same route (single-path routing) or over different routes (multipath-routing) [6].

In order to adjust in a flexible mode, the capacity allocated to connections, the network must be capable of dynamically changing the size of the containers in a hitless manner, *i.e.* without affecting the service. The resizing of ODUflex containers can be accomplished using the protocol Hitless Adjustment of ODUflex (HAO), while the members of a VCG can be added or removed through the Link Capacity Adjustment Scheme (LCAS). Both techniques have their own advantages and drawbacks. ODUflex is easier to implement and manage than VCAT and, as each signal is transported as a single entity, it does not require differential delay compensation, as it is required with the second technique. However, resizing operations are more complex for ODUflex paths than for VCAT ones, since they require the participation of all nodes in the path, contrary to VCAT, where only the ingress and egress nodes take action in the operation. Furthermore, when multipath routing is provided, the scheme based on Virtual Concatenation permits to implement traffic engineering techniques, such as load balancing, guaranteeing the use of network resources more efficiently [7]. In addition, LCAS can be employed for resilience purposes [8]-[11]: it can automatically remove disrupted VCG members in the presence of link failures, assuring that an unprotected ODU connection still continues operating despite working at a lower capacity; it can also be employed to activate backup VCG members used in protected connections whenever necessary. The first scheme is particular useful in data communications with unprotected connections where it is preferable to have a connection working at lower bit rate with "degraded service", rather than no connection at all. The second scheme rely on the existence of backup VCG members, which are set up in advance using paths which are link disjoint from the working ones, to protect the working members.

In traffic engineering applications, the reconfiguration time of a VCG is not a crucial issue as far as the operation does not take place in real time. However, in protection applications, it is important to be able to calculate the time required for adding or removing VCG members from a connection in order to compute the fault recovery time, *i.e.* the time elapsed between the instant a failure is detected and the instant the traffic is recovered.

This paper focus on the problem of VCG reconfiguration in OTN networks by using LCAS and details how the reconfiguration time can be calculated considering some typical reference networks. Although the impact of LCAS on the dynamic bandwidth adjustment in the context of Next Generation (NG)-SDH/SONET networks has been previously analyzed [12], no similar analyses have been published on OTN networks to the best of our knowledge. Furthermore, we address the problem of evaluating the fault recovery time in the cases where the VCAT/LCAS is also applied for resilience purposes.

The rest of this paper is organized as follows. Section 2 reviews the operating principles of VCAT and LCAS technologies. Section 3 explains how the time required by LCAS to add or remove VCG members, as well as the fault recovery time, can be calculated. Section 4 adds some illustrative examples considering well-known reference networks and Section 5 concludes the paper.

2. VCAT and LCAS Overview

The OTN is designed to accommodate different client signals both at wavelength and sub-wavelength granularity [13]. The sub-wavelength operation is based on an electrical layered structure comprising the Optical Channel Payload Unit (OPU), Optical Channel Data Unit (ODU) and Optical Transport Data Unit (OTU). The OTU layer is the electrical content of the Optical Channel (OCh), which itself is the basic unit to be used when wavelength granularity is required. The VCAT in the OTN is realized by logically aggregating X OPUk (k = 1, 2, 3) signals. Note that Virtual Concatenation for OPUk with k = 0, 2e, 4, flex is not supported by the standard. The aggregated signal corresponds to the VCG and is denoted as OPUk-Xv, where X is in the range from 1 to 256 and the lowercase v denotes Virtual Concatenation. The structure of the OPUk-Xv frame is depicted in **Figure 1** using a bi-dimensional representation.

It consists of a matrix of octets with 4 rows and X×3810 columns, where columns 14X+1 to 16X correspond to the OPUk overhead area and columns 16X+1 to 3824X to the payload area (OPUk-Xv column numbers are derived from the OPUk columns in the ODUk frame). The columns 14X+1 to 15X include the Virtual Concatenation Overhead (VCOH) formed by the three octets VCOH1/2/3, which are used to carry the control information responsible for the VCAT process. It can also be referred that columns 1 to 14X, which are omitted in **Figure 1**, correspond to the ODUk and OTUk overhead and, as a consequence, are only included in the OTN frame structure. The capacities of the different VCGs in an OTN network are shown in **Table 1** [6].

The presented results reveal that VCAT applied in OTN networks can achieve quite impressive capacities, much far beyond the ones that ODUflex can offer, since those are limited by the ODUs capabilities. Note, for example, that it will be possible to transport in the future a flow of about 10 Tb/s using an OPU3-Xv, while without VCAT the use of an OPU3 only allows the transport of about 40 Gb/s.

The implementation of VCAT requires the usage of control signals which reside mainly on the OPU overhead. Two control signals are defined: the Multi-Frame Indicator (MFI) and the Sequence Number (SQ). The MFI is used to numerate each successive payload container frames of the traffic flow, in such a way that all OPUks of the same VCG have the same MFI. On the other hand, each OPUk member of a VCG has its own and unique SQ, which is in the range from 0 to (X−1). The sequence numbers are used by the destination node to reconstruct the original payload containers sequence having, in a certain way, a similar role to the sequence numbers used in the Real Time Protocol [14].

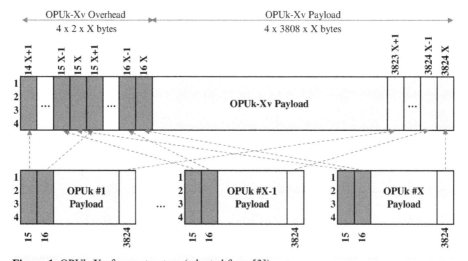

Figure 1. OPUk-Xv frame structure (adapted from [3]).

Table 1. Capacities for different VCGs with VCAT.

VCG type	X range	Capacity (Gb/s)
OPU1-Xv	1 to 256	~2.488 to ~637.010
OPU2-Xv	1 to 256	~9.995 to ~2558.710
OPU3-Xv	1 to 256	~40.151 to ~10278.533

As the VCAT employs a two-stage multi-frame, there is one MFI per stage. The MFI of the first stage uses the Multiframe Alignment Signal (MFAS) of the OTN frame alignment overhead area as an 8-bit indicator, and cycles from 0 to 255. As the MFAS is incremented by 1 every OPUk frame, the first stage multiframe (MFAS multiframe) has 256 OPUk frames.

The MFI of the second stage includes the MFI1 and MFI2 bytes to form a 16-bit indicator, which cycles from 0 to 65,535 since it is incremented at the start of each MFAS multiframe (MFAS = 0), thus it can take 65,536 different values. The bytes MFI1 and MFI2 are located in the first and second octet, respectively, of the Virtual Concatenation Overhead (VCOH1), while the bytes SQ are placed in the fifth octet, as shown in **Figure 2**.

The LCAS permits to change dynamically the size of a VCG by adding or removing members in a hitless manner with the operation being controlled by a network management plane, or by a control plane like GMPLS [15], or even by a network operating system using the OpenFlow protocol [16]. LCAS is implemented using a number of control signals, which also reside in the VCOH1 octet of the OPU overhead, with exception of the Member Status (MST) field which resides in the VCOH2 octet. From the source node to destination node, besides the MFI and SQ, LCAS also uses the Control (CTRL) word and the Group Identification (GID) bit. In the opposite direction, *i.e.* from the destination node to the source node, LCAS uses the MST field and the Re-Sequence Acknowledge (RS-Ack) bit. The CTRL field has the following states:

- FIXED: the number of members of the concatenated group cannot be changed (VCAT without LCAS);
- ADD: this member is going to be added to the concatenation group;
- NORM: this member is active and is used to transport data;
- EOS (End-Of-Sequence): this member is the last of the concatenation group;
- IDLE: this member is not part of the concatenation group or is in the process of being removed;
- DNU (Do-Not-Use): this member has a failed path to the destination node and must not be used. The backup members used to protect working members, while not active must be in this state.

The MST field is used to report the status of all the VCG members from the destination node back to the source node, using for that purpose a multi-frame defined by the last five bits of the MFAS signal (MST multi-frame), which are used to form a 5-bit indicator, which cycles from 0 to 31. The status of each member has two states:

- OK: This member is part of the concatenation group and has been correctly received at the destination;
- FAIL: This member is not part of the concatenation group, or has been received with failures.

In addition, GID identifies the VCG with all members of the same group having the same GID value, while RS-Ack is used by the destination to inform the source that it is aware of a change in SQ sequence of the VCG members.

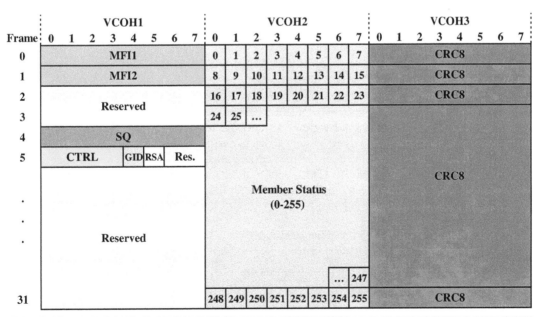

Figure 2. OPUk-Xv virtual concatenation overhead (adapted from [3]).

3. Time Delay of LCAS Operations

3.1. Components of the Time Delay

The time delay introduced by LCAS operations includes the multiframes propagation delay, the node processing delay and the LCAS message processing time. The first two terms are determined by the network physical topology. Assuming that a given path between a source and a destination node has a length l and passes through n intermediate nodes, the contribution of these terms is described by:

$$t_d = l \times \tau_f + n \times \tau_n, \tag{1}$$

where τ_f is the propagation delay per km and τ_n is the latency per node. For a typical optical fiber $\tau_f = 5$ μs/km and the maximum latency per node is $\tau_n = 25$ μs [9]. The LCAS message processing time depends on the operations being performed. In the following lines two cases will be analyzed: 1) The addiction of a new member to a VCG, leading to an increase of the connection capacity; 2) The removal of an existent member from a VCG, leading to a decrease of the connection capacity. In both cases the processing time of LCAS depends on the time required to generate an MFAS multiframe (t_{MFAS}), the time to generate an MST multiframe (t_{MST}) and the number of multiframes exchanged to complete the adjustment. The presented results include, besides the LCAS message processing time, its propagation delay. The node processing delays are omitted for simplicity since they are negligible regarding the overall operation delays.

In OTN networks the duration of MFAS and MST multiframes is given by the duration of a single OPUk frame (t_{OPUk}) multiplied by the number of frames corresponding to the structure, which is 256 and 32, respectively. The duration of these multiframes is shown in **Table 2**. The MST multiframe requires 32 frames to transmit the state of all the members of the VCG OPUk-256v, because this information is transmitted in the byte VCOH2 of the OPUk overhead and, as a result, a single frame can only transmit the state of 8 VCG members.

3.2. Connection Capacity Increase

The process of increasing the capacity of a connection by adding new members to a given VCG (OPUk-Xv in OTN) is assumed to be initiated by the Network Management System (NMS). As shown in **Figure 3**, in an initial state the new member to be added has its CTRL command set to IDLE, since it is not yet a member of the VCG, its SQ number set to the maximum value supported and its MST value set to FAIL. In order to increase the connection capacity the NMS sends a request to the source node for the addition of a new member to the VCG. As a result the following operations take place:

1) The source node generates and transmits an MFAS multiframe, where the CTRL word of the member to be added to the VCG is changed from IDLE to ADD, as an indication to the destination node that the correspondent member is going to be added, together with the new assigned SQ (higher than the previous maximum used SQ). This corresponds to the time to generate and transmit an MFAS multiframe plus the time of its propagation to the destination node $(2t_{MFAS} + t_d)$.

2) After the destination node having received the CTRL = ADD it sends back an MST multiframe with the status of the new member changed to OK. Thus the delay of this action is the time needed to transmit an MST multiframe plus the time of its propagation to the source node $(t_{MST} + t_d)$.

3) Once the source receives this acknowledgment it generates and transmits a new MFAS multiframe with the new member added. The new member has now its CTRL set to EOS and its SQ assigned, so the decoder can know the correct order of the members when they arrive to destination node. This action takes the time to generate and transmit an MFAS multiframe plus the time of its propagation to the destination node $(2t_{MFAS} + t_d)$. In this analysis it is considered that if a new CTRL code is to be inserted while a given MFAS multiframe is being transmitted, the new control code can only be inserted after finishing the multiframe transmission, whence it is used $(2t_{MFAS})$ in the calculations [12].

Table 2. MFAS and MST multiframe duration.

	OPU1	OPU2	OPU3
t_{OPUk}	48.971 μs	12.191 μs	3.035 μs
t_{MFAS}	12.537 ms	3.121 ms	0.777 ms
t_{MST}	1.567 ms	390.112 μs	97.120 μs

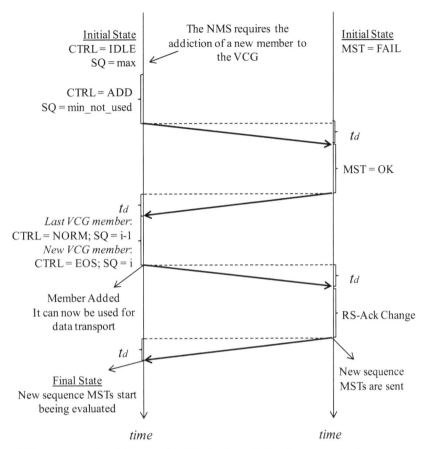

Figure 3. Time diagram for LCAS capacity increase process.

4) When the destination node receives the new member sequence, with the new member added, it sends back the RS-Ack bit changed, in an MFAS multiframe, as an indication that now it has knowledge of the new sequence. The time needed for this operation is the time of transmitting the MFAS multiframe plus the time of its propagation to the source node $(t_{\mathrm{MFAS}}+t_d)$. The addiction of the new member to the OPUk-Xv is then concluded.

Hence, the overall time delay corresponding to the capacity increase operation is given by:

$$D_{\mathrm{add}} = 5t_{\mathrm{MFAS}} + t_{\mathrm{MST}} + 4t_d. \tag{2}$$

Neglecting the propagation and the node delay the total delay time to add a new member to a VCG is then 64.252 ms for OPU1, 15.995 ms for OPU2 and 3.982 ms for OPU3.

3.3. Connection Capacity Decrease

In an initial state the member to be removed from the OPUk-Xv has its CTRL set to NORM or EOS (if it is the member of the VCG with highest SQ), its SQ set to "i", and its MST set to OK. In order to decrease the connection capacity the NMS sends a request to the source node for the removal of that member from the VCG. As a result the following operations (see **Figure 4**) take place:

1) The source node generates and transmits an MFAS multiframe with the CTRL word of the member to be removed changed from NORM or EOS to IDLE. Besides, if the member to be removed is the last member of the VCG, the CTRL word of the previous member is changed from NORM to EOS and its SQ is kept unchanged. If the member to be removed is not the last member of the VCG, the SQ numbers of all following active members are decremented by one and their CTRL words are kept unchanged. This corresponds to the time of generating and transmitting an MFAS multiframe plus the time it takes to reach the destination node $(2t_{\mathrm{MFAS}}+t_d)$.

2) After receiving CTRL = IDLE the destination sends an MST multiframe with the status of the member to

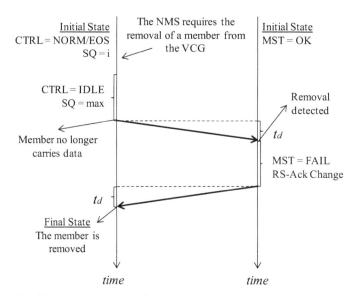

Figure 4. Time diagram for LCAS capacity decrease process.

be removed as FAIL, as an indication of its acknowledgment of the member removal, and the RS-Ack bit changed for that member, as an acknowledgment of the new sequence in the VCG. Thus the delay of this action is the time needed to transmit an MST multiframe plus the time of its propagation to the source node $(t_{MST} + t_d)$. The removal of the member from the VCG is then concluded.

Thus, the overall capacity decrease operation delay is given by:

$$D_{rem} = 2t_{MFAS} + t_{MST} + 2t_d.$$ (3)

As a consequence, the total time required to remove a VCG member, when the propagation and the node delay are neglected, is 26.641 ms for the OPU1, 6.632 ms for the OPU2 and 1.651 ms for the OPU3.

3.4. Fault Recovery Time

In the presence of a network failure the first step of the recovery process consist in removing the failed member from the VCG. This process involves the following steps (see **Figure 5**):

1) The destination node detects a failure in a working VCG member in time instant $t_f (t_f \leq t_d)$.

2) The destination node removes the failed member from the payload reassembly process and reports the failure to the source by changing its status to MST = FAIL. The delay of this action is then the time needed to transmit an MST multiframe plus the time required to reach the source node $(t_{MST} + t_d)$. Note that for a certain period of time the re-assembled payload in the destination side will be harmed, since the traffic is still sent by the source in all the pre-fault members of the VCG.

3) When the source node receives MST = FAIL, it notifies the NMS about the detected failure and generates and transmits a new MFAS multiframe with the CTRL word changed to DNU and at the same time stops putting data on the payload area of the failed member. Once the code CTRL = DNU arrives to the destination node, the removal process is complete, and the payload of the VCG is now error free. Thus the delay of this action is the time required to generate and transmit an MFAS multiframe plus its propagation time $(2t_{MST} + t_d)$ As a consequence the maximum fault-recovery time can be described as:

$$D_{rec} \leq 2t_{MFAS} + t_{MST} + 4t_d.$$ (4)

Neglecting the contribution of the time required to detect the failure $(\leq t_d)$, one concludes that the recovery time given by (4) is exactly the same as the time required to the remove a VCG member (3).

In protection schemes based on the "degraded service", the fault recovery time is given by Equation (4). However, in protection schemes that rely on the existence of protection/backup resources the calculation is different. These schemes require the pre-provision of additional capacity by adding backup members to the VCG in addition to the working members, as a way to protect the last ones. The backup members do not carry any traffic

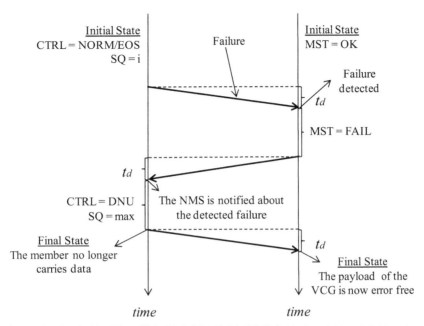

Figure 5. Time diagram for LCAS failed member removal.

during normal operation and to guarantee that they are not used by the destination side in the reassembly process their CTRL status is set to be DNU. Therefore, the fault recovery time in this case, besides the time required for detecting and notifying the failure of a working member, also include the time to activate the backup capacity. The first time includes the steps 1) and 2) of the previously described recovery process, while activating the backup members involves the following actions (see **Figure 6**).

3) After receiving the notification of a member failure the source node notifies the NMS about it and changes the CTRL field of the backup member from DNU to EOS or NORMAL, and the CTRL field of the failed member from EOS or NORMAL to DNU. The SQ numbers are also rearranged. This new coding takes the time $(2t_{MFAS} + t_d)$ to reach the destination side.

4) The destination node detects CTRL = EOS (or NORMAL) for the backup member and consequently will start to transmit MST = OK. Remember that the failed member is already transmitting MST = FAIL. The time taken by this action is $(t_{MST} + t_d)$. Once the source receives MST = OK the traffic previously transmitted on the failed member is switched to the backup member and the recovery process ends. Assuming that the time required by the source to switch the traffic is negligible, the fault recovery time of the described protection scheme reduces to:

$$D_{pro} \leq 2t_{MFAS} + 2t_{MST} + 4t_d.$$ (5)

when the propagation and node delay are not taken into account the fault recovery time is 28.208 ms for the OPU1, 7.022 ms for the OPU2 and 1.748 ms for OPU3.

The ITU-T Recommendation G.841 [17] indicates that in NG-SDH/SONET networks based on ring protection, for a ring with a perimeter of less than 1,200 km, the switching completing time for a single failure must be less than 50 ms. For a distance of 1,200 km, the propagation time is 6 ms. In this case, neglecting the node latency, we conclude that the worst case fault recovery time is 52.208 ms for the OPU1, 31.022 ms for the OPU2 and 25.748 ms for OPU3, showing that for the OPU2 and OPU3 the values are well below the requisite of the typical value of 50 ms.

4. Simulation Results

The methodology presented in the previous section is used here to evaluate the time delays of the LCAS operations in different OTN networks. For sake of comparison, the time delays for NG-SDH/SONET networks are also evaluated using the results presented in [12]. To obtain the propagation delays the shortest-path between each source-destination node pair was computed using the Dijkstra algorithm.

In our analysis, we considered three network topologies: **Figure 7(a)** the 24-node North American backbone network (UBN), which has 42 bidirectional links and all links are shorter than 3,000 km, **Figure 7(b)** the 19-node European Optical Network (EON) with 36 bidirectional links, and the longest link is about 2,000 km, and **Figure 7(c)** the Pan-European test network (COST 239), which comprises 11 nodes and 26 bidirectional links, and all links are shorter than 1,000 km (the number on each link represents the length in km). In our study it was computed the maximum and the mean delay related to LCAS operations in each network for both NG-SDH/SONET and OTN technologies. The maximum LCAS delays were computed using the shortest path between the two farthest network nodes, while the mean LCAS delays require the knowledge of the mean value of the shortest-paths computed between all network node pairs.

The results obtained for the time delay introduced by LCAS are shown in **Figure 8(a)** for the UBN network, **Figure 8(b)** for the EON network and **Figure 8(c)** for COST 239network. It was considered the scenarios where connection capacity increases, connection capacity decreases, and a protection switching is performed between a failed working member and a backup member. For the NG-SDH/SONET networks both the Low Order VCAT

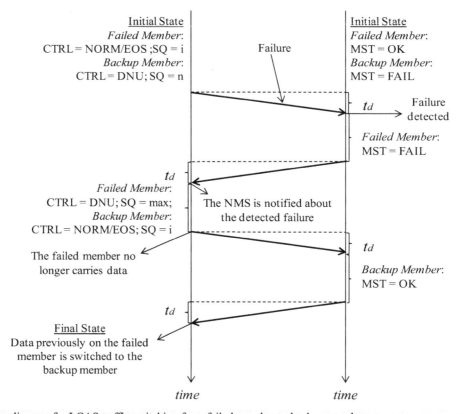

Figure 6. Time diagram for LCAS traffic switching from failed member to backup member.

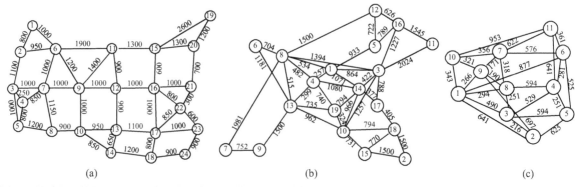

| (a) | (b) | (c) |

Figure 7. Physical topology of (a) UBN; (b) EON and (c) COST 239 network.

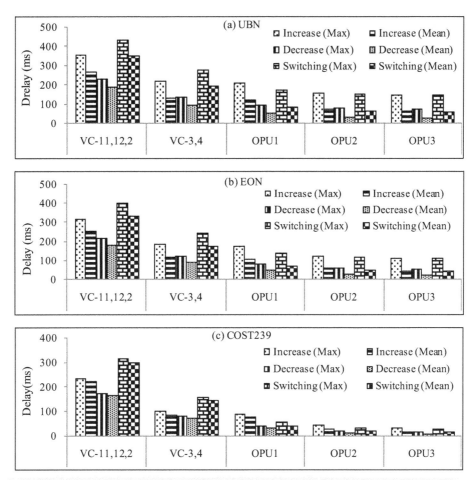

Figure 8. LCAS delays in (a) UBN; (b) EON and (c) COST 239 network.

(LO VCAT) for VC-11, VC-12 and VC-2 concatenated containers, and High Order VCAT (HO VCAT), for VC-3 and VC-4 concatenated containers [18] are considered. **Table 3** emphasizes the maximum LCAS time delays for capacity increase and protection switching, since they are the most critical ones.

The most obvious result is the significant difference between time delays in NG-SDH/SONET networks for LO VCAT and HO VCAT signals, being the latter greatly smaller. This difference comes from the fact that the duration of LO VCAT and HO VCAT frames is 500 µs and 125 µs, respectively [18], which impacts the duration of the multiframes used in the LCAS analysis.

In OTN networks this delay is even more reduced, since the frame durations are smaller than in NG-SDH/SONET networks. Thus, this becomes immediately an advantage of OTN technology.

In OTN networks the impact of the propagation delay of the messages exchanged between the source and destination nodes is more relevant than it is in NG-SDH/SONET networks, since the OPUk frames duration is significantly smaller than the VC-n frames duration. With k = 1 the delay of increasing or decreasing an OPU1-Xv connection capacity is similar to the delay of increasing or decreasing, respectively, the capacity of an HO VC-n-Xv[1] connection. However, for higher values of k the delay of LCAS operations decrease substantially, since the multiframe duration also decreases. Regarding to the operation of switching traffic from a failed member to a previously provisioned backup one, the fact of using an OPU1-Xv instead of an HO VC-n-Xv leads to a delay reduction of approximately half.

In all the analyzed scenarios the operation of increasing the link capacity takes more time than the inverse operation, because the handshaking procedure between the source and the destination nodes requires more steps. Furthermore, the difference between the time delays corresponding to these two operations is larger in OTN networks. This is due to the fact that the duration of the MFAS multiframe impacts more the first operation than

[1]Notation of a VCAT VCG, where n identifies the virtually concatenated containers type.

Table 3. Maximum LCAS time delay for capacity increase and protection switching (ms).

		Capacity Increase Delay (ms)			Protection Switching Delay (ms)		
Technology	Container	UBN	EON	COST 239	UBN	EON	COST 239
NG-SDH	LO VCAT	352	317	236	432	397	316
	HO VCAT	218	183	102	276	241	160
	OPU1	208	173	92	172	137	56
OTN	OPU2	160	125	44	151	116	35
	OPU3	148	113	32	146	111	29

the second one (see Equations (2) and (3)) and this duration is much larger than the duration of the MST multi-frame, contrary to what happens in the NG-SDH/SONET networks, where the MST multiframe is longer than the multiframe used for the MFI control [18]. For the same reason, the operation of protection switching from one member to another is quite large in NG-SDH/SONET since the destination node transmits two MST multi-frames to the source node. As for OTN, this operation's delay is similar to the capacity increase delay, since the propagation time has a major impact in the overall operation delay and the number of multiframes exchanged in both cases is equal. Note, for example, that in COST 239 network we can accomplish with an OPU3-Xv a delay of 29 ms, while in a NG-SDH/SONET networks we get no less than 160 ms.

The multiframe delays are constant for each LCAS operation and each VC-n-Xv or OPUk-Xv. Therefore, delays suffered during the process of dynamically allocating or freeing bandwidth vary with the distance between network nodes. Hence, LCAS operation delays are network topology dependent.

The UBN network, which presents longer path distances, naturally suffers bigger delays and, as a consequence, the contribution of the propagation delay to the LCAS delay is stronger than, for example, in the COST 239 network. The maximum distance between two nodes in the latter network is 1,386 km. This leads to a maximum propagation delay around 36 ms, for increasing the connection capacity. For the UBN network the maximum distance between network nodes goes up to 7,200 km leading to a maximum propagation delay of about 144 ms.

5. Conclusion

In this paper, we have explored the application of VCAT/LCAS techniques to provide dynamism in the context of OTN networks. A detailed explanation about the procedures used to resize the capacity of the connections is presented and the time-delays associated with the process are computed. A comparison with NG-SDH/SONET networks is also provided. It is shown that the resizing operations in OTN networks are faster than in NG-SDH/SONET networks and the speed of the process increases when we move from OPU1-Xv connections to OPU3-Xv connections. For example, for the first type of connection the maximum time delay obtained in all the reference networks considered was about 200 ms, while for the second one the maximum delay is reduced to about 150 ms. For the sake of comparison, the worst results for NG-SDH/SONET networks were about 220 ms and 350 ms, for HO-VCAT and LO-VCAT, respectively. Our results have also highlighted the interest of applying the VCAT/LCAS techniques as a way to improve resilience: by adding a backup member to protect a working member in a VCG we showed that using OPU3-Xv connections it is possible to recover from a member failure in about 25 ms, in a scenario where the NG-SDH/SONET standards define a maximum value of 50 ms.

References

[1]   Gumaste, A. and Krishnaswamy, N. (2010) Proliferation of the Optical Transport Network: A Use Case Based Study. *IEEE Communications Magazine*, **48**, 54-61. http://dx.doi.org/10.1109/MCOM.2010.5560587

[2]   Carrol, M., Roese, J. and Ohara, T. (2010) The Operator's View of OTN Evolution. *IEEE Communications Magazine*, **48**, 46-52. http://dx.doi.org/10.1109/MCOM.2010.5560586

[3]   ITU-T Rec. G.709 (2009) Interfaces for the Optical Transport Network (OTN).

[4]   ITU-T Rec. G.7044/Y.1347 (2011) Hitless Adjustment of ODUflex (GFP).

[5]   ITU-T Rec. G.7042/Y.1305 (2006) Link Capacity Adjustment Scheme (LCAS) for Virtual Concatenated Signals.

[6]   Bernstein, G., Caviglia, D., Rabbat, R. and Van Helvoort, H. (2006) VCAT/LCAS in a Clamshell. *IEEE Communications Magazine*, **44**, 34-36. http://dx.doi.org/10.1109/MCOM.2006.1637944

[7]   Santos, J., Pedro, J., Monteiro, P. and Pires, J. (2011) Optimized Routing and Buffer Design for Optical Transport Networks Based on Virtual Concatenation. *Journal of Optical Communications and Networking*, **3**, 725-738. http://dx.doi.org/10.1364/JOCN.3.000725

[8]   Acharya, S., Gupta, B., Risbood, P. and Srivastava, A. (2004) PESO: Low Overhead Protection for Ethernet over SONET Transport. *IEEE Infocom* 2004, Hong Kong. http://dx.doi.org/10.1109/INFCOM.2004.1354491

[9]   Roy, R. and Mukherjee B. (2008) Degraded-Service-Aware Multipath Provisioning in Telecom Mesh Networks. *OFC/NFOEC* 2008, San Diego, 24-28 February 2008. http://dx.doi.org/10.1109/OFC.2008.4528661

[10] Huang, S., Martel, C. and Mukherjee, B. (2011) Survivable Multipath Provisioning with Differential Delay Constraint in Telecom Mesh Networks. *IEEE/ACM Transactions on Networking*, **19**, 657-669. http://dx.doi.org/10.1109/TNET.2010.2082560

[11] Ou, C., Sahasrabuddhe, L., Zhu, K., Martel, C. and Mukherjee, B. (2006) Survivable Virtual Concatenation for Data over SONET/SDH in Optical Transport Networks. *IEEE/ACM Transactions on Networking*, **14**, 218-231. http://dx.doi.org/10.1109/TNET.2005.863462

[12] Han, D., Li, X. and Gu, W. (2006) The Impact of LCAS Dynamic Bandwidth Adjustment on SDH/SONET Network. *Journal of Optical Communications*, **27**, 317-320. http://dx.doi.org/10.1515/JOC.2006.27.6.317

[13] Pedro, J., Santos, J. and Pires, J. (2011) Performance Evaluation of Integrated OTN/DWDM Networks with Single-Stage Multiplexing of Optical Channel Data Units. *ICTON* 2011, Stockholm, 26-30 June 2011. http://dx.doi.org/10.1109/ICTON.2011.5970940

[14] Schulzrinne, H., Casner, S., Frederick, R. and Jacobson, V. (2003) RTP: A Transport Protocol for Real-Time Applications. *RFC* 3550.

[15] Bernstein, G., Caviglia, D., Rabbat, R. and van Helvoort, H. (2011) Operating Virtual Concatenation (VCAT) and the Link Capacity Adjustment Scheme (LCAS) with Generalized Multi-Protocol Label Label Switching (GMPLS). *RFC* 6344.

[16] Das, S., Parulkar, G., Singh, P., Getachew, D., Ong, L. and McKeown, N. (2010) Packet and Circuit Network Convergence with Open Flow. *OFC/NFOEC*'10, San Diego, 24-28 March 2010.

[17] ITU-T Rec. G.841 (1998) Types and Characteristics of NG-SDH Network Protection Architectures.

[18] Van Helvoort, H. (2005) Next Generation NG-SDH/SONET: Evolution or Revolution? John Wiley & Sons, Ltd., Chichester. http://dx.doi.org/10.1002/0470091223

12

A Review Study of Wireless Sensor Networks and Its Security

Muhammad Umar Aftab[1,5], Omair Ashraf[2], Muhammad Irfan[3], Muhammad Majid[4], Amna Nisar[5], Muhammad Asif Habib[5]

[1]Department of Computer Science and Engineering, Yuan Ze University, Taiwan
[2]Department of Computer Science, NFC-IEFR, Faisalabad, Pakistan
[3]Land Record Management and Information System, Faisalabad, Pakistan
[4]IT Department, Post Graduate College, Faisalabad, Pakistan
[5]Department of Computer Science, National Textile University, Faisalabad, Pakistan
Email: ms.umaraftab@yahoo.com, omairashraf@outlook.com, multi_com51@hotmail.com, mmajid2026@gmail.com, nisar390@yahoo.com, dr.m.asif.habib@gmail.com

Abstract

The confluence of cheap wireless communication, sensing and computation has produced a new group of smart devices and by using thousands of these kind of devices in self-organizing networks has formed a new technology that is called wireless sensor networks (WSNs). WSNs use sensor nodes that placed in open areas or in public places and with a huge number that creates many problems for the researchers and network designer, for giving an appropriate design for the wireless network. The problems are security, routing of data and processing of large amount of data etc. This paper describes the types of WSNs and the possible solutions for tackling the listed problems and solution of many other problems. This paper will deliver the knowledge about the WSN and types with literature review so that a person can get more knowledge about this emerging field.

Keywords

WSN, Overview of WSN, WSN with Types, WSN Security, RBAC

1. Introduction

WSN has become an emerging field in research and development due to the large number of applications that can become significantly beneficial from such systems and has led to the development of cost effective, not-reusable, tiny, cheap and self-contained battery powered computers, also called sensor nodes. These sensor nodes can accept input from an attached sensor and process the input data gathered from the sensor nodes. After that the process data wirelessly transmits the results to transit network. WSNs are highly dispersed networks of lightweight and small wireless nodes, deployed in huge numbers, to monitor the system or environment by the measurement of physical parameters like pressure, temperature, or relative humidity [1]. China put intelligent information processing and sensor network in priority for 15 years in the "National medium and long term program for science & development (2006-2020)". WSNs can be applied in industry, agriculture, military defense, environment monitoring, remote control and city management etc. that is why WSNs are becoming more and more popular [2] [3].

WSNs have much more similarity with Mobile Ad-hoc Networks (MANET). WSNs also create network that contains sensor nodes connecting with each other, in an Ad-hoc manner and no proper infrastructure is there for both but WSNs have the collection of data with the sensor nodes but MANET can or cannot use sensor nodes. In this paper, we gave the description of WSNs and its types with literature review, as shown in the **Figure 1**. WSNs consist of tiny and low power sensor nodes that collect data through tiny sensors, process the data and send to particular location. We also describe the types of WSNs with the research work. We include the flaws of existing technology or in a particular type and how we can cover those open holes by using various techniques, protocols or algorithms.

2. Types of WSN

2.1. Mobile Wireless Sensor Networks (MWSNs)

MWSNs can be defined as a WSN that have mobile sensor nodes as compared to the usually used WSN in which sensor nodes are static. MWSNs have more versatility than the static WSNs because MWSNs can be deployed for any scenario and they can manage with quick topology changes.

The normal WSN is simply deployed with static nodes to achieve monitoring missions in the area of interest but due to dynamic changes of hostile environment and events, a pure static WSN may face the following problems:

- Connectivity of the whole network and complete coverage of the sensing area could not possible in WSN like in the case of robots or aircrafts for hostile region [4].
- As sensor nodes usually works with battery powered and prone to errors. The node can be dead if the energy of battery ends and this results the communication breakup of sensor network and replacement of new nodes is also a difficult task.
- For some special applications like tracking applications, the network needs a larger nodes to cover the whole area that ultimately the cost of network is increased.
- For some applications, there is a need of some sophisticated sensors for performing some specific military tasks that may need camera with every sensor node for image collection that is not feasible to equip every node with separate camera.

By introducing mobility, all the listed problems can be overcome and many other problems can be covered. We can enhance the flexibility and capability of WSN by adding mobile nodes. Different missions can be conduct by controlling the movement of mobile sensors [5]. More and more, individuals and communities are using

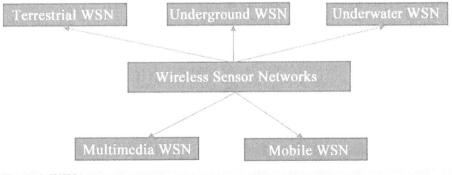

Figure 1. WSN types.

MWSNs and there is no need of pre-deployed network infrastructure when cooperating mobile nodes communicate with each other, in order to achieve various kinds of functions [6].

MWSNs are deployed in an open environment and having more chances of security attacks. Security is one of key issue in MWSNs and need to be solved. The attacks in MWSNs can occur from any side or any direction to any targeted node because MWSNs are consists of mobile wireless nodes that forms the temporary network without any centralized infrastructure [7] [8]. The complex security mechanisms and algorithms cannot be implemented in wireless mobile sensor nodes. The reason of not implementing any security algorithm or mechanism is due to resource constraints regarding bandwidth, computational power and memory size. The traditional security mechanisms are invalid for MWSNs due to the mobility of wireless nodes in network topology and this mobility creates dynamic attributes in topology that results invalidity of security mechanisms. The security attacks (internal or external) can be controlled or minimized by using the cryptography and authentication mechanisms but both techniques can handle the external attacks in WSNs and are unable to handle the internal attacks in MWSNs because the wireless sensor nodes can be easily stolen when deployed in hostile or in an open environment. For this case, the network can be controlled or destroyed by the nodes which have accessed the network [9].

One of the attack in WSN is the node replication attack and number of protocol proposed for tackling these type of attacks but no suitable mechanism is find out for MWSNs. However an appropriate mechanism is described by Deng, Xiong and Chen [10]. They described the mobility property and propose two protocols for mobility assistance for the detection of node replication attacks in MWSNs. First protocol is the Unary Time Location Storage and Exchange (UTLSE) that assigns each observer a task of tracking a particular set of other nodes. All the observers only store one time location entitlement for every tracked node and detect the replication when they come across each other. The second protocol is Multi Time Location Storage and Diffusion (MTLSD) that lets every observer stores multi time location claims for each tracked node. It also introduces more cooperation between the observers to improve the detection performance. Both protocols works as encounter-based because they only sent messages for detecting the replication when two nodes meet or come across each other and due to this way of working, the protocols do not have any need of routing signaling messages. Both protocols can also identify the replication with high detection accuracy as well as with very low communication, computation and storage overheads [10].

A three layer architecture for mobile nodes is described, in which all the sensors in MWSNs are organized, by using the architecture. Data collection, data processing and routing table maintenance are placed in different nodes. The complexity of sensors and the cost of construction can be reduced by using Multi-layer MWSNs (M2WSN). SP (Shortest path) routing protocols proposed on the base of new architecture, for adapting sensors to update the network topology. The researchers also include a simulation of SP that reduces the energy utilization and offers a decent solution for node movement in M2WSNs and this simulation shows the better results, as compared to LEACH [11].

2.2. Underwater Wireless Sensor Networks (UWSNs)

Underwater wireless communication is one of the major challenge in building UWSN. It has been observed that Radio Frequencies and acoustic waves (having narrow bandwidth) are heavily attenuated and altered in water. An alternative but a feasible solution that can be considered is using optical communication, in case of short range distance. This approach mainly emphasizes on an Optical Physical (PHY) Layer taking into account the features of WLAN (IEEE 802.11) Infrared Physical Layer and the compatibility with the most recent terrestrial Wireless Sensor Network's protocol *i.e.* IEEE 802.15.5. As compared to acoustic communication, if optical communication in green/blue wavelengths (for short distances) are used then they offer high band communication and faster propagation in water. An experimental set up was done and it was noticed that increasing the distance (between a LED and a photodiode) causes a high BER (Bit Error Rate) while water turbidity was also kept in mind [12].

Terrestrial wireless sensor networks are an active area for development and research. Fundamental properties of these networks are low-power, a several number of co-operating small nodes that are capable in observing, detecting, and tracking various objects and events inside a specific environment. This makes these networks very appealing for a number of military and industrial applications. Within the overall wireless sensor network field, underwater wireless sensor networks (UWSN) is an emerging area of research. The number of underwater wireless sensor networks-based applications is continually increasing. Most of the UWSN applications can be classified as Seismic, monitoring, assisted control and navigation, location reference points and security applications. While future UWSN includes applications for attack purposes and unmanned submarines. When it comes to UWSN, there are still so many challenges and difficulties to deal with e.g. power-consumption, security, and time synchronization, communication between UWSNs, installation, implementation, power recharging and recovery [13].

2.3. Space-Based Wireless Sensor Networks (SB-WSNs)

The wireless sensor networks are networks of integrated micro sensors for monitoring and data gathering for some of the environment conditions *i.e.* temperature, vibration, sound, motion and pressure. While in space, these networks might be used for space weather purposes in (LEO) low Earth Orbit or implementation of wireless sensor networks within a spacecraft in single probe missions or in order to interchange electrical wires, or as very tiny satellite (sensor) nodes flying in compact formations and chemical and physical sensing of the soils, surfaces and atmospheres of other planets. Multipath routing scheme is a perfect nominee for space-based missions of micro-sensor nodes. WSNs need to be optimized if they are to be used for space or solar system exploration. The modifications should be according to space requirements. Design issues like selection and design of antenna, software and power supply must be completed by thoroughly examining the mission's characteristics [14].

The concept of terrestrial wireless sensor networks (TWSN) can be applied to space *i.e.* satellite sensor network. Grouping the design and enabling technologies for pico-satellite formations. The idea is to use inexpensive constellation of sensor nodes to collect important information instead of doing the same using a large expensive satellite. The research that's been carried out at the Surrey Space Centre, was mainly aimed at space weather missions in low Earth orbit. Future space-crafts are thought to be miniature, autonomous and distributed. In this regard, flower constellation set is considered to be the best for orbital configuration in nano- and micro-satellite missions. While there are some issues in Flower constellation, related to positioning satellites which can be solved using Inter-satellite communication capability. Communication issues, referring to the Open Systems Interconnection (OSI) networking scheme, of a space based wireless sensor network (SB-WSN) have been summarized. A system-on-a-chip computing model and platform and the agent middleware for SB-WSNs have been presented. This system architecture focused on the LEON3 soft processor core is targeted at effective hardware support of collaborative processing in these networks, offering several intellectual property cores *i.e.* transceiver core, a hardware accelerated Wi-Fi MAC and a Java co-processor. A new configurable inter-satellite communications component for pico-satellites has also been outlined [15].

2.4. Wireless Underground Sensor Networks (WUSNs)

The probabilistic connectivity of the WUSNs has been discussed. WUSNs are one of the unique extension of terrestrial WSNs. WUSNs' heterogeneous network architecture and channel characteristics, the connectivity study is much more complicated than in the ad hoc networks and terrestrial WSNs. This connectivity issue might haven't been addressed previously. Thus, a mathematical model was developed to study and examine the probabilistic connectivity in WUSNs, which gathered the effects of environmental parameters *i.e.* the soil composition and soil moisture, and several system parameters *i.e.* the sensor burial depth, the operating frequency, the density of the sensor devices, the sink antenna height, the number and the mobility of the above-ground sinks and the tolerable latency of the networks. The upper and lower bounds for the connectivity probability are calculated systematically. Simulation and investigation studies were performed, whereas the theoretical bounds were authenticated, and the effects of system parameters and some environmental parameters on the performance were explored [16].

2.5. Wireless Multimedia Sensor Networks (WMSNs)

The Wireless Multimedia Sensor Networks (WMSNs) comprise of tiny sensor-nodes that can sense, compute, actuate, communicate, and have control components. Various applications of the Wireless Sensor Multimedia Networks (WMNs) include target trailing, habitation monitoring, traffic management systems and ecological monitoring; these kinds of applications involve efficient communication of event happenings and features in multimedia form *i.e.* image, audio and video [17].

Wireless Multimedia Sensor Network (WMSN) is a novel appliance of Wireless Sensor Networks (WSNs), as Multimedia data needs continuous transmission of data, increased bandwidth, storage and power, and low latency rate so WMSNs requires much attention. So far different routing protocols have been proposed for proficient data communication in WSNs. Usually in WSNs, the routing algorithms designed to route tiny scalar data for comparatively short time interval. The basic ingredients of WSNs routing protocol are use of minimum hops, maximize the available power, achieve low latency rate and less load of traffic, finding more than one path to destination etc. With sensor networks another significant concern is the creation of Holes which is because of the fact that during routing, the nodes nearby the destination are used more frequently so in result the batteries of such nodes gets exhausted in advance. Thus such nodes failed to transmit the sensor information to the base station [18].

Hop and Load based Energy Aware Routing protocol (HLEAR) has been developed for WMSNs, to eradicate the above described issues. In HLEAR, the algorithm finds semi disjoint or disjoint paths by hop counts, load of traffic and energy of nodes. As HLEAR is a reactive (On Demand) routing protocol so a compression is made between HLEAR and Tiny Advanced On-demand Distance Vector (AODV), a reactive protocol. The HLEAR protocol found more intelligent in path selection as it chooses such paths which can carry affording traffic rate by having lesser hop distance. Furthermore, due to absorption of energy of nodes, HLEAR also tackles with Hole creation problem [18].

In WMSNs, we mostly comes up with a question that "How and where to Install sensor nodes while having inadequate resources of communication to support all nodes?" Researchers tried to answer this question by making wide theoretical analysis. Its general considerations are less mobile or a static type of networks environment, smooth topology and TDMA based single channel communication to show cross layer design model in which both node admittance to a WMSNs mechanism and the interaction of node with the resource management and link scheduling mechanisms, are examined. Generally interaction of node is originated with two stage optimization problem in which the first step is to increase the total numbers of already acknowledged sensor nodes and the second step is to enlarge the lifetime of the network. The interaction of node can also originates as one stage optimization problem having more complex mathematical logic. The commonality among all described proposals is the segregation of some sort of services (shown in **Figure 2**), without this segregation the QoS cannot be guaranteed by the WMSNs [19].

As described earlier that WMSNs requires much more recourses like as bandwidth. Spectrum sensing approach is used to reply the request of bandwidth by which the spectrum utilization maximizes. Hence to utilize the spectrum holes and available bands there is a need of reliable, accurate, efficient and real time methods. Previously, in cognitive radio (CR) some methods are introduced to sense the spectrum, among those methods the Multi-Taper Method (MTM) is the most tempting. As MTM is considered very near optimal for wideband signals and also an efficient method for CR so we can assume that for spectrum sensing in WMSNs, MTM can be a superior selection. The existing MTM have some challenges like supplementary resource demand. Therefore, introduced MTM into WMSNs and also present an algorithm to eradicate the previous implementation issues in MTM. The detailed simulations proved that the new approach for wideband signals in WMSNs is more corresponding to the real time values and also provides much less false alarm rate as compare to other methods due to decreased variance. This approach also detects the spectrum holes more efficiently and accurately [20].

For the purpose of optimization of network performance introduced an Energy Efficiency QoS Assurance Routing in Wireless Multimedia Sensor Networks (EEQAR). EEQAR is actually an analysis of social network to improve performance of network. The main idea behind the development of EEQAR is to introduce such

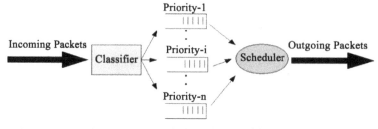

Figure 2. Segregation of services [19].

routing for wireless multimedia sensor networks which provides energy efficient assurance of QoS. For the selection of most consistent paths, link quality estimators are not used in EEQAR, though it generates an additional load on EEQAR in the process of route discovery to communicate between different clusters. The video quality levels evaluation does not covered by EEQAR [21].

The researchers suggest the concept that for multiple path communication among two nodes, Multipath Data Transfer protocol offers concurrent multiple paths. Their proposed algorithm divides the work between all nodes which equally extends the overall life of WMSNs [17].

2.6. Terrestrial Wireless Sensor Networks (TWSNs)

Most generally the Terrestrial WSNs contains hundreds to thousands of cheap wireless sensor nodes which are installed in a specified geographical area. The deployment can be in an ad-hoc network or in pre-planned networks based. In the case of Ad-Hoc networks, the sensor nodes can be released from plane and arbitrarily place them into the area of target. In the case of pre-planned, there are four different placements as followed, Grid, Optimal, 2-D and 3-D placement models [22].

The FSO/RF systems in wireless sensor networks are getting much attraction from researcher. A free-space optical(FSO) link used in FSO/RF systems as basic communication medium while a RF (Radio Frequency) links are also used as backup when LOS for optical communication are not present. As FSO optical communication links results low communication energy than the high data rate broadband optical communications, so the idea of using FSO links in WSNs get heap. The major concern of FSO/RF is weather effects like snow or rain. For terrestrial applications, the comparison of the lifetime performance of hybrid WSNs and FSO/RF WSNs under weather effects of rain and snow shows that by proper threshold selection we can achieve the most favorable practice of power efficient FSO link [23].

2.7. WSN Security and Security Issues

Generally WSNs are used to collect information from various locations of physical world and also they are deployed in controlled and uncontrolled environment [3]. So by their applications and deployment nature Wireless sensor networks are ultimately insecure. These networks have numerous limitations like node (less computational power, less memory, less energy etc.), network (because they are acting as mobile as hoc network) and physical (deployed in different environments like public and hostile) limitations which makes them supplementary vulnerable to various security attacks. Ad hoc nature of sensor networks opens the unique challenges to the reliability and security. Owing to the limited computational and processing constrains traditional security techniques and policies are not suitable in order to maintain confidentiality, Authentication, Availability and Integrity in WSN, s [1].

According to Pfleeger there are four different classes of security threads that are common in computational systems and also in sensor networks [24]. These are Interruption, Interception, Modification and Fabrication. In Wireless sensor networks researcher identified several possible security attacks like passive information gathering, node subversion, false node, node malfunction, node outage, message corruption, traffic analysis, routing loops, selective forwarding, sinkhole, Sybil, wormholes, hello flood and DoS etc. These attacks also disturbs WSN layers specifically application, transport, network, data link and physical layers. Different countermeasures and defense techniques are presented by researchers for layered security like malicious node detection and isolation, unique pair wise keys for application layer, limiting connections numbers, client puzzles for transport layer, Key management secure routing, Authentication, Encryption, Redundancy, Probing, monitoring, two way and three authentication and three way handshake for network layer, link layer encryption, rate limitations, error correcting code for data link layer and adaptive antennas, spread spectrum for physical layer. However we need to have a security framework in order to provide countermeasures against security attacks in WSN [24] [25].

Wireless sensor networks (WSNs) are extremely prone and susceptible to external and internal attacks as they consist of numerous devices with constraints for example; less memory, associated low energy and low battery power. The nodes in WSNs communicate with each other through wireless links. Nevertheless, WSNs are being deployed extensively. There are still unsolved problems in WSNs and security is one of the high priority research issues. These networks are implemented in hostile environments. Resource-constraints, communication overheads involved, and environmental conditions give rise to various security attacks or threats. Securely communication and security among WSN nodes is an important challenge. Authors described the security of

WSNs and attacks that occur at different layered architecture of WSNs and how to prevent them [26].

Secure protocols should be designed and some access control mechanism can be applied to provide a secure network for mobile devices in an organization. Different researchers work on the security and access control mechanisms. Role Based Access Control (RBAC) is one of the most widely used access control model. The main thing in RBAC is the management of large number of permissions with help of Roles. RBAC architectural issues in institution collaborative systems has been highlighted because there is no specific architectural design that had been defined for the institution's security. A system has been proposed that combines the efficiency of both RBAC and Organization Unit (OU). The objective of using OU for the institutional systems is the effective management of users and objects. Proposed system helps in two things: one user management and secondly load sharing of administrator. By applying OU, users can be easily be managed on department level and as well as in a particular department. Also the OU creation would be done on department level. Furthermore, a hierarchical architectural model has been described that can help an institution or IT manager in implementing and deploying of RBAC with the concept of OU thus, making a system more secure and facilitating the users in efficient manner [27].

3. Conclusion and Future Work

Networks are shifted from wired to wireless quickly but wireless networks are costly but in wireless networks; WSNs is growing day by day and hot field in the area of research. WSNs are cost effective because it saves the energy by using low power tiny sensor nodes that makes it popular, with the addition of different other features. WSNs have a variety of features and types that can accommodate many problems arising in different scenarios. The only need is the selection of the right approach on the right place, for getting the maximum benefit from the WSN and its types. We have a plan to find out an algorithm or mechanism that improves the performance and security issues, of the WSN. This paper enhances the base for this emerging field and after it we will pick a particular problem in WSN and work for an efficient approach.

References

[1]   Sharma, K. and Ghose, M. (2010) Wireless Sensor Networks: An Overview on Its Security Threats. *IJCA*, *Special Issue on "Mobile Ad-hoc Networks" MANETs*, 42-45.

[2]   Li, J.Z. and Hong, G. (2008) Survey on Sensor Network Research. *Journal of Computer Research and Development*, **45**, 1-15.

[3]   Ren, F.Y., Lin, G. and Huang, H.N. (2003) Wireless Sensor Networks. *Journal of Software*, **7**.

[4]   Dhillon, S.S. and Chakrabarty, K. (2003) Sensor Placement for Effective Coverage and Surveillance in Distributed Sensor Networks. *IEEE*, **3**, 1609-1614.

[5]   Rezazadeh, J. (2012) Mobile Wireless Sensor Networks Overview. *International Journal of Computer Communications and Networks* (*IJCCN*), **2**, 17-22.

[6]   Ren, Y. and Boukerche, A. (2008) Modeling and Managing the Trust for Wireless and Mobile Ad Hoc Networks. *IEEE International Conference on Communications*, 2008. *ICC'08*, Beijing, 19-23 May 2008, 2129-2133. http://dx.doi.org/10.1109/icc.2008.408

[7]   Cho, J.-H., Swami, A. and Chen, R. (2011) A Survey on Trust Management for Mobile Ad Hoc Networks. *IEEE Communications Surveys & Tutorials*, **13**, 562-583. http://dx.doi.org/10.1109/SURV.2011.092110.00088

[8]   Djenouri, D., Khelladi, L. and Badache, N. (2005) A Survey of Security Issues in Mobile Ad Hoc Networks. *IEEE Communications Surveys*, **7**, 2-28.

[9]   Duan, J., Qin, Y., Zhang, S., Zheng, T. and Zhang, H. (2011) Issues of Trust Management for Mobile Wireless Sensor Networks. *7th International Conference on Wireless Communications*, *Networking and Mobile Computing* (*WiCOM*), Wuhan, 23-25 September 2011, 1-4.

[10]   Deng, X.M., Xiong, Y. and Chen, D.P. (2010) Mobility-Assisted Detection of the Replication Attacks in Mobile Wireless Sensor Networks. *IEEE 6th International Conference on Wireless and Mobile Computing, Networking and Communications* (*WiMob*), Niagara Falls, 11-13 October 2010, 225-232.

[11]   Duan, Z.-F., Guo, F., Deng, M.-X. and Yu, M. (2009) Shortest Path Routing Protocol for Multi-Layer Mobile Wireless Sensor Networks. *International Conference on Networks Security*, *Wireless Communications and Trusted Computing*, *NSWCTC'09*, Wuhan, 25-26 April 2009, 106-110. http://dx.doi.org/10.1109/NSWCTC.2009.282

[12]   Anguita, D., Brizzolara, D. and Parodi, G. (2009) Building an Underwater Wireless Sensor Network Based on Optical:

Communication: Research Challenges and Current Results. *3rd International Conference on Sensor Technologies and Applications*, *SENSORCOMM'* 09, Athens, 18-23 June 2009, 476-479. http://dx.doi.org/10.1109/SENSORCOMM.2009.79

[13] Davis, A. and Chang, H. (2012) Underwater Wireless Sensor Networks. *Oceans*, 2012, Hampton Roads, 14-19 October 2012, 1-5. http://dx.doi.org/10.1109/oceans.2012.6405141

[14] Akbulut, A., Patlar, F., Zaim, A. and Yilmaz, G. (2011) Wireless Sensor Networks for Space and Solar-System Missions. 2011 *5th International Conference on Recent Advances in Space Technologies* (*RAST*), Istanbul, 9-11 June 2011, 616-618. http://dx.doi.org/10.1109/RAST.2011.5966912

[15] Vladimirova, T., Bridges, C.P., Paul, J.R., Malik, S.A. and Sweeting, M.N. (2010) Space-Based Wireless Sensor Networks: Design Issues. *IEEE Aerospace Conference*, Big Sky, 6-13 March 2010, 1-14.

[16] Sun, Z. and Akyildiz, I.F. (2010) Connectivity in Wireless Underground Sensor Networks. 2010 *7th Annual IEEE Communications Society Conference on Sensor Mesh and Ad Hoc Communications and Networks* (*SECON*), Boston, 21-25 June 2010, 1-9. http://dx.doi.org/10.1109/secon.2010.5508264

[17] Poojary, S. and Pai, M.M. (2010) Multipath Data Transfer in Wireless Multimedia Sensor Network. 2010 *International Conference on Broadband*, *Wireless Computing, Communication and Applications* (*BWCCA*), Fukuoka, 4-6 November 2010, 379-383. http://dx.doi.org/10.1109/BWCCA.2010.100

[18] Nayyar, A., Bashir, F. and Hamid, Z. (2011) Intelligent Routing Protocol for Multimedia Sensor Networks. 2011 *International Conference on Information Technology and Multimedia* (*ICIM*), Kuala Lumpur, 14-16 November 2011, 1-6. http://dx.doi.org/10.1109/icimu.2011.6122747

[19] Phan, K.T., Fan, R., Jiang, H., Vorobyov, S.A. and Tellambura, C. (2009) Network Lifetime Maximization with Node Admission in Wireless Multimedia Sensor Networks. *IEEE Transactions on Vehicular Technology*, **58**, 3640-3646. http://dx.doi.org/10.1109/TVT.2009.2013235

[20] Shafiee, M. and Vakili, V. (2012) MTM-Based Spectrum Sensing in Cognitive Wireless Multimedia Sensor Networks (C-WMSNs). 2012 *6th International Symposium on Telecommunications* (*IST*), Tehran, 6-8 November 2012, 266-270. http://dx.doi.org/10.1109/ISTEL.2012.6482995

[21] Lin, K., Rodrigues, J.J., Ge, H.W., Xiong, N.X. and Liang, X.D. (2011) Energy Efficiency QoS Assurance Routing in Wireless Multimedia Sensor Networks. *IEEE Systems Journal*, **5**, 495-505. http://dx.doi.org/10.1109/JSYST.2011.2165599

[22] Akyildiz, I.F., Su, W.L., Sankarasubramaniam, Y. and Cayirci, E. (2002) A Survey on Sensor Networks. *IEEE Communications Magazine*, **40**, 102-114. http://dx.doi.org/10.1109/MCOM.2002.1024422

[23] Nadeem, F., Leitgeb, E., Awan, M. and Chessa, S. (2009) Comparing the Life Time of Terrestrial Wireless Sensor Networks by Employing Hybrid FSO/RF and Only RF Access Networks. *5th International Conference on Wireless and Mobile Communications*, *ICWMC'* 09, Cannes, 23-29 August 2009, 134-139. http://dx.doi.org/10.1109/ICWMC.2009.29

[24] Pfleeger, C.P. and Pfleeger, S.L. (2003) Security in Computing. Prentice Hall Professional, Upper Saddle River.

[25] Zia, T. and Zomaya, A. (2006) Security Issues in Wireless Sensor Networks. *International Conference on Systems and Networks Communications*, *ICSNC'* 06, Tahiti, 29 October-3 November 2006, 40-40. http://dx.doi.org/10.1109/ICSNC.2006.66

[26] Singh, S. and Verma, H.K. (2011) Security for Wireless Sensor Network. *International Journal on Computer Science & Engineering*, **3**, 2393-2399.

[27] Aftab, M.U., Nisar, A., Asif, M., Ashraf, A. and Gill, B. (2013) RBAC Architectural Design Issues in Institutions Collaborative Environment. *International Journal of Computer Science Issues* (*IJCSI*), **10**, 216-221.

13

Effective Life and Area Based Data Storing and Deployment in Vehicular Ad-Hoc Networks

Hirokazu Miura[1,2], Hideki Tode[2], Hirokazu Taki[1]

[1]Faculty of Systems Engineering, Wakayama University, Wakayama, Japan
[2]Department of Computer Science and Intelligent Systems, Osaka Prefecture University, Sakai, Japan
Email: miurah@sys.wakayama-u.ac.jp

Abstract

In vehicular ad-hoc networks (VANETs), store-carry-forward approach may be used for data sharing, where moving vehicles carry and exchange data when they go by each other. In this approach, storage resource in a vehicle is generally limited. Therefore, attributes of data that have to be stored in vehicles are an important factor in order to efficiently distribute desired data. In VANETs, there are different types of data which depend on the time and location. Such kind of data cannot be deployed adequately to the requesting vehicles only by popularity-based rule. In this paper, we propose a data distribution method that takes into account the effective life and area in addition to popularity of data. Our extensive simulation results demonstrate drastic improvements on acquisition performance of the time and area specific data.

Keywords

Vehicular Ad-Hoc Networks (VANETs), Popularity, Effective Life, Effective Area, Data Deployment

1. Introduction

In an Intelligent Transportation System (ITS) [1], a number of applications for safety, comfort and convenience have been proposed. Many of them rely on distributing data, e.g., on the current traffic conditions [2], or on free parking spaces around the current location of the vehicle [3]. Entertainment and information services such as multimedia communication and messaging or advertising from roadside to vehicle are also very attractive applications [4]. In ITS, information sharing based on inter-vehicle communication is effective for improving data availability. ITS involves two categories of communication, vehicle-to-infrastructure and vehicle-to-vehicle communication. In vehicle-to-infrastructure communication, a distributed database stored at fixed sites, such as roadside units, is queried by the moving vehicles via the wireless network infrastructure. Then, the vehicle obtains data from the database and carries it in the direction of travel. In vehicle-to-vehicle communication, two vehicles can communicate with each other when their distance is smaller than a wireless communication range which can be connected by a local area wireless protocol such as IEEE 802.11 [5], Bluetooth [6], and so on. These protocols provide broadband but short-range peer-to-peer communication, and enable vehicular ad-hoc networks (VANETs) [7]. In VANETs, a mobile user discovers the desired information from the vehicles it encounters or from distant vehicles by multi-hop transmission relayed by intermediate moving vehicles [8]. This direct communication between individual vehicles can significantly increase passenger comfort. These are so called application layer store-carry-forward approach and categorized in Disruption Tolerant Networks (DTNs) [9].

In the above approach, as the amount of data items carried by a vehicle becomes larger, the performance such as the ratio of desired data reception is more improved. This is because a vehicle has many chances that it can encounter another vehicle which has desired data. Storage resource in a vehicle, in other words, the amount of saved data is limited in general. Therefore, attributes of data that should be carried by vehicles are an important factor in order to efficiently disseminate desired data. So far, there has been Roadcast study as a typical popularity aware content sharing scheme in VANETs. Roadcast consists of two components called popularity aware content retrieval and popularity aware data replacement. The popularity aware content retrieval scheme finds the most relevant and popular data for user's query. The popularity aware data replacement ensures that different data is deployed inside a vehicular network according to its popularity. Roadcast achieves that more popular data tends to be shared with other vehicles so that the query delay and the query hit ratio can be improved. The existing methods such as Roadcast deploy the data to vehicles randomly according to its relative popularity. In VANETs, however, there are different types of data which depend on the time and location, and some information has an effective life or a deployment area. Such kind of data cannot be deployed adequately to the requesting vehicles only by popularity-based rule, and the data out of effective scope may not be a high valuable for requesting user even if it is obtained. Existing method based on popularity cannot achieve the system which takes into account the effective scope of the data. Therefore, in this paper, we propose a data distribution method that takes into account the effective life and area in addition to popularity of data.

The rest of this paper is organized as follows. Related work is discussed in Section 2. Section 3 describes our system model. Section 4 presents the effective life and area based data storing and deployment. Performance evaluations are shown in Section 5. Finally, we conclude the paper in Section 6.

2. Related Works

2.1. VANET

Vehicular networks represent an interesting application scenario not only for traffic safety and efficiency but also for more commercial and entertainment support. So far, however, most of vehicular network researches focus on routing issues [3] [7] [10]. They all assume the consumer related information is known beforehand so that the sender can route the content to its destination. For example, VADD studies how to choose the best routing path based on the traffic information. Other researches in vehicular networks have focused on content distribution [7] [11] [12]. The efficient discovery and distribution of information is a challenging problem especially in a dynamic environment such as vehicular network. Literature [7] introduces data pouring and buffering techniques to disseminate data along the roads. This paper studies content sharing, where each vehicle queries useful data from its encountered neighbor vehicles. Different from destination aware routing and dissemination [7], how to disseminate the most suitable data to neighboring vehicles is the main focus of this paper.

In the last couple of years there has been an increasing interest in in-network aggregation mechanisms for vehicular ad hoc networks [13]. This technology aims at reducing redundant information and improving communication efficiency by summarizing information that is exchanged between vehicles. Our deployment takes into consideration the number of replicas so that data with high density in the network is not stored in vehicle's storage, but not summarizing information.

On the other hand, content retrieval through intermittent contact opportunities in vehicular networks is also an important technique. In literature [4], content retrieval is studied in a small area, where vehicles in adjacent lanes exchange information as they pass through one another. The scheduling issues of content retrieval at the road intersection are analyzed in [14]. To accelerate content retrieval, randomized network coding is proposed [11]. However, the network coding based data diffusion brings in large amount of redundant data which may not be useful but taking much communication bandwidth and memory space. In Roadcast, content retrieval [15] is based on users query request and how to efficiently share content with future encountered vehicles based on local information is studied.

Studies in data replacement start from cache replacement. In literature [16], several replacement algorithms for web cache are studied. Later, these replacements are improved by adding the popularity factor. However, all these works are based on web cache which is in a centralized environment. Roadcast differs from the existing works in that it is a distributed replacement algorithm and it aims to optimize the network-wide content sharing performance.

2.2. Roadcast

Roadcast [15] has been proposed as a system for sharing information on VANET. In Roadcast, popular data is distributed to many vehicles so that it can satisfy many users request in the future. Roadcast achieves these objectives with two techniques. One is popularity aware content retrieval and the other is popularity aware data replacement. First, the popularity aware content retrieval scheme makes use of information retrieval (IR) techniques to find the relevant data towards user's queries. However, different from the traditional IR techniques, the factor of data popularity is considered and the relevance of the data to queries is re-ranked, so that more popular data is more likely to be shared with other vehicles. Second, in Roadcast, the downloaded data is stored as replica which can be shared with other Roadcast users. When the local memory is full, some data objects have to be replaced. The proposed data replacement algorithm ensures that the data replications with different popularity can have different life time so that popular data can have more, while not too many, copies in the network. Roadcast considers the popularity of data as a most important factor to improve the query hit ratio.

2.3. Issues of Roadcast

In Roadcast, the popularity of data is considered as the most important factor. However, effective life and area of the content is not taken into account. In general, various kinds of information are shared in VANET. This information include not only entertainment information such as MP3 music or video but also restaurant and parking information, sale advertisement of shops located on the roadside, and so on. Such information may be delivered only for specific areas or may have limited valid time. In other words, the value of the information may be reduced or become invalidout side effective life and area. When such content is disseminated, not only popularity of the content but also effective life and area should be taken into account. In the paper, we propose the data deployment based on the effective life and area as well as popularity of data.

3. System Model

In vehicular ad-hoc networks (VANETs), moving vehicles carry data and exchange it as they pass each other. In this section, we describe our system model. In our system, a vehicle obtains data from the wireless network infrastructure. A vehicle with obtained data moves on a road and encounters another vehicle on the opposite lane. Then the vehicle exchanges the obtained data each other. The vehicle obtains the desired data by repeating this behavior. **Figure 1** shows the system model.

In **Figure 1**, vehicle A moves to the right side and vehicle B moves to the left side, respectively. When the vehicle A passes near the data source, the data source deploy data in the vehicle. Then, the vehicle A moves and carries the obtained data. If the vehicle A encounters the vehicle B which requested the data in the vehicle A, the data is exchanged between A and B.

4. Effective Life and Area Based Data Storing and Deployment

4.1. Replacement Algorithm

In vehicular ad-hoc networks (VANETs), moving vehicles carry data and exchange it as they pass each other. Storage resource in a vehicle, in other words, the amount of saved data is limited in general. Therefore, attributes of data that should be carried by vehicles are an important factor in order to efficiently disseminate de-

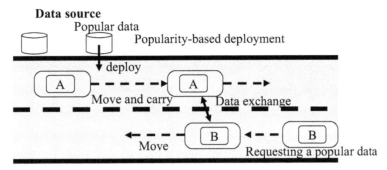

Figure 1. System model.

sired data. In this paper, we focus on effective life and area as well as popularity of data, and propose a replacement algorithm based on these attributes. The proposed method decides the data to be replaced according to the following operations.

In this method, the data which has the largest w_i, defined as Equation (1), has to be replaced from the storage of the vehicle, if new data is input when the storage capacity is full.

$$w_i = \frac{1}{f_i} * \left\{ \alpha g_i + (1 - \alpha) c_i \right\} * T_i * D_i \tag{1}$$

$$(0 \le \alpha \le 1)$$

where f_i, g_i and c_i are popularity, generation number of data and the replication count of data i, respectively. T_i and D_i are calculated based on effective life and area of its data.

Popular data should not be replaced if it has not been disseminated yet, in order to give more opportunities to disseminate the data with higher popularity. In our method, therefore, each data source decides whether it deploys a specific object or not with a probability which is predefined by its popularity. When the total number of vehicles is V and the request probability of data i is P_i, the expected number of vehicles which obtain data i is VP_i. This means that the data is deployed randomly in a VANET so that the number of replicated data is linear to its popularity. Furthermore, in our method, it can be expected that data with low popularity is also deployed to vehicles because of the probabilistic manner. In our system, it is also necessary to collect popularity information of all data which is expected to be requested from users.

The generation number g_i corresponds to the number of vehicles which are transferred from its original source. The replication count c_i represents the number of copies which are generated by the same vehicle. If the data has large g_i or c_i, it can be expected that many copies of the data exist in the VANET. Therefore, this system is willing to replace the data which has higher g_i and c_i to avoid deployment of redundant copies and store other kinds of data. α is the coefficient that decides which g_i or c_i is more important.

T_i represents a temporal effectiveness of data i and is obtained by normalizing the age of data i to adjust the scale of the other elements. T_i is calculated by the following equation.

$$T_i = \text{elapsed time from the data } i \text{ generation time} / \text{effective life time of data } i$$

If $T_i < 1$, the age of data i is within the effective life, otherwise the data i grow stale.

D_i indicates a spatial effectiveness of data i, similarly with temporal effectiveness. D_i is expressed by the following.

$$D_i = \text{distance from original location of data } i / \text{radius of the effective area for data } i$$

There are different possible types of functions for T_i and D_i function. In this paper, we use above function as one example in which the effectiveness decays linearly with time and distance.

4.2. Scheduling Method of Data Deployment

In VANET where vehicles move at high speed, since the time for the data exchange is less, the amount of data which can be transmitted and received at a time is also limited. Thus, the performance of our system depends on the scheduling method of data deployment. In the paper, we proposed the following scheduling method.

When a vehicle encounters the other vehicle, our proposal prefers to deploy the data which has the smallest w_i, defined as Equation (1). It can be expected that popular data within effective life time and area is distributed faster.

5. Performance Evaluation

In this section we describe the simulation model and compared model for our system evaluation, and then present the simulation results.

5.1. Simulation Setup

To investigate the performance of our deployment, we evaluate probability of data reception, *i.e.*, the percentage of data items which requesting vehicles successfully obtained. In our evaluation, we implement our deployment on the ONE simulator (ver.1.4.0) [17] and use a map of the Helsinki area (**Figure 2**). There are totally 100 roadside units on this map. A data object is generated by each roadside unit. When the effective life of the data is expired, the roadside unit generates a new data object. Some keywords are assigned to each data object or each user's query, according to Zipf-like distribution [18]. The simulation parameters are summarized in **Table 1**.

5.2. Compared Method

In this evaluation, we compare the performance of the proposed method with popularity-only which decides the data to be deployed according to its popularity. In the popularity-only, the data which has the largest w_i', defined in Equation (2), has to be replaced from the storage of the vehicle if new data is input when the storage capacity is full. We assume this method works as conventional method.

$$w_i' = \frac{1}{f_i} * \{\alpha g_i + (1-\alpha)c_i\}$$

$$(0 \le \alpha \le 1)$$

(2)

5.3. Simulation Results

Figure 3 shows the query hit ratio with several size of buffer memory. The query hit ratio is the possibility that the query can be served by local vehicle or roadside unit. **Figure 4** shows the query delay with several size of buffer memory. The query delay is defined as the average delay from initiating the query to receiving the re-

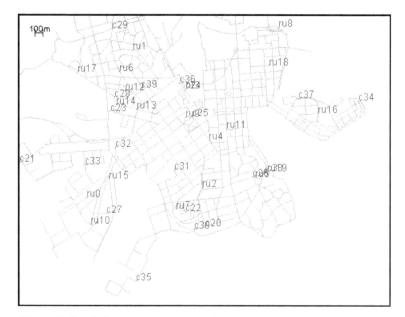

Figure 2. Simulation map.

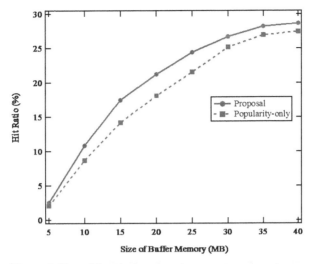

Figure 3. Query hit ratio.

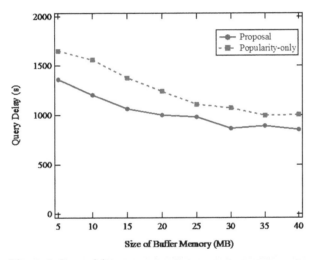

Figure 4. Query delay.

quired data. From **Figure 3**, our proposal consistently outperforms popularity-only in query hit ratio, the improvement is up to around 3%. As shown in **Figure 4**, when the memory size is small, both schemes have a relatively higher query delay. When the memory size increases, the query delay decreases. This is because as the memory size increases, vehicles are able to buffer more data objects. Hence, there will be more data replicas and the queries can be served by these replicas quickly. From these figures, we also observe our proposal slightly outperforms popularity-only.

Furthermore, we also evaluate the performance of our proposal in terms of effective data. **Figure 5** shows query hit ratio of the effective data which means percentage of the data within its effective life and area in successfully obtained data. **Figure 6** shows its query delay. From these results, we can observe that our proposal significantly improved these performances on valid data acquisition.

Figure 7 shows the query hit ratio with different Zipf parameters. On the Internet, popularities of data follow Zipf-like distribution. This distribution indicates that the number of data with high popularity is small and many data have low popularity. When the Zipf parameter is large, the number of data with low popularity is large and the popularity of few data with high popularity becomes enhanced. Therefore, Popularity-only shows better performance than our proposal. However, when the Zipf parameter is small, access pattern tends to be quite uniformly distributed, and different keywords have similar popularity. From **Figure 7**, when the content access is close to uniform distribution, our proposal has much advantage.

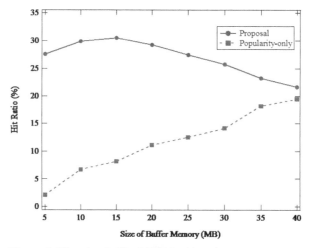

Figure 5. Hit ratio of effective data.

Table 1. Simulation parameters and their values.

Parameter	Value
Simulation Time	12 h
Number of Vehicles	150
Simulation Area	4.5 km × 3.4 km (Helsinki area)
Communication Range	200 m
TransmissionSpeed	250 KBps
Data Size	500 KB - 2 MB
Size of Buffer Memory	5 MB - 40 MB
Keyword Set Size	40
Number of Keywords in Data	2 - 8
Number of Keywords in Query	3 - 5
Effective Life of Data	0.5 h - 3 h
Effective Range of Data	500 m - 2 km
Zipf Parameter	0.5
Vehicle Speed	10 m/s - 15 m/s
α value	0.5

5.4. Effective Data Ratio in the Effective Area

We also evaluate effective data ratio in its effective area according to elapsed time from data generation. We set the parameters as shown in **Table 2** and the other parameters are same as shown in **Table 1**.

In **Figure 8**, x-axis shows the elapsed time from data generation and y-axis shows the ratio of the vehicles which have data in the vehicles in effective area. In this case, effective life of data is 1h. In the popularity-only, even if the effective life is expired, the amount of data in effective areas increases and ineffective data is still deployed. After a while, the deployed ineffective data starts to decrease. On the contrary, in our proposal, although effective data ratio starts to decline before the expiration of effective life, more valid data is distributed to more vehicles in effective area. **Figure 9** and **Figure 10** are results in the case where effective life is 0.5 h and 1.5 h, respectively. These figures show the similar results as **Figure 8**. From these results, it is possible for our

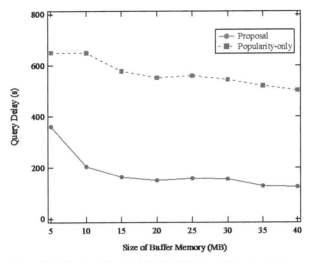

Figure 6. Query delay of effective data.

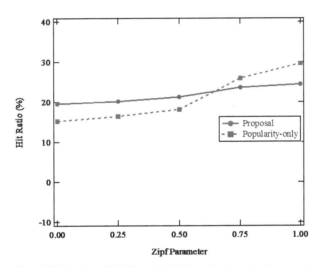

Figure 7. Impact of Zipf parameter.

Table 2. Simulation parameters.

Parameter	Value
Size of Buffer Memory	15 MB
Effective Range of Data	500 m
Effective Life of Data	0.5 h, 1 h, 1.5 h

proposal to distribute more valid data to the effective area within the effective life of data. In the proposed method, data is gradually deleted from vehicles' buffer before its expiration. Thus, our proposal can suppress that the vehicle buffer is occupied by invalid data, and release the useless buffer resources of the vehicle for other more valid data. Therefore, area-limited information can be distributed to most effective vehicles by our proposal method.

5.5. Impact of Replacement Policy

In this section, we compare the performance on each factor of replacement policy. **Figure 11** represents the result of data acquisition performance in the case of popularity only, popularity and effective area, popularity and

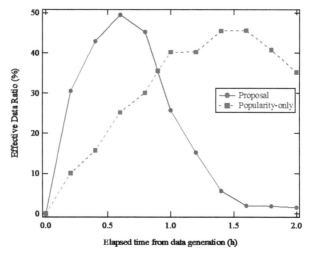

Figure 8. Effective data ratio (Effective life: 1.0 h).

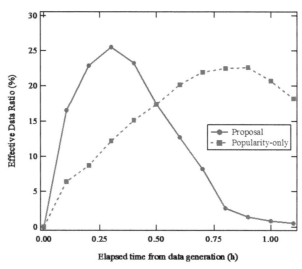

Figure 9. Effective data ratio (Effective life: 0.5 h).

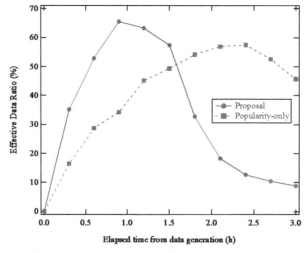

Figure 10. Effective data ratio (Effective life: 1.5 h).

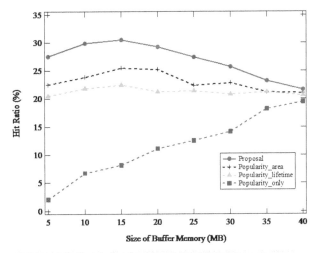

Figure 11. Impact of replacement algorithms.

effective life, and considering all factors (proposal). As the size of buffer memory becomes large, there is no difference in the buffered data, and all policy shows similar performance. However, in the case where the size of buffer memory is small, it is possible for the proposed method to consider the expiration time and scope of data at the same time, and achieve performance improvement.

5.6. Impact of Parameter α

In this section, data acquisition performance is evaluated when a parameter α in Equation (1) is changed. **Figure 12** shows the result in the case of 100, 150 and 200 vehicles. From this result, by setting α to around 0.3 to 0.4, the performance can be improved. Therefore, it is important to balance between the number of copies and the number of generations to obtain a better performance.

5.7. Impact of Deployment Scheduling Method

In VANET where vehicles move at high speed, since the time for the data exchange is less, the amount of data which can be transmitted and received at a time is also limited. In this section, in order to investigate the differences in performance of the scheduling of deployment, we evaluate the performance of our proposal in comparison with the following scheduling method. In this evaluation, we have adopted our proposed replacement to both scheduling methods.

(FIFO) The data to be transmitted is decided by FIFO. Neither the expiration time nor scope of the data is taken into account. The data is transmitted according to the order of old data which the vehicle stored.

The above method (FIFO) is evaluated in the same manner as **Figure 5** and the result is shown in **Figure 13**. By our proposed scheduling, distributing the locally valid data that depends on the time and place shows better performance. The effect of the scheduling method depends on the time for data exchange. It is also expected that this effect can increase as the number of transmitted data at a time is small.

6. Conclusions

In this paper, we propose a deployment method which can help a user get the useful data as much as possible through intermittently connected VANET. In VANETs, store-carry-forward approach may be used for data sharing, where moving vehicles carry and exchange data when they go by each other. In this approach, attributes of data that have to be stored in vehicles are an important factor in order to efficiently distribute desired data. In VANETs, there are different types of data which depend on the time and location. Such kind of data cannot be deployed adequately to the requesting vehicles only by popularity-based rule, and the data out of effective scope, such as effective life and area, may not be a high valuable for requesting user even if it is obtained. Existing method based on popularity cannot achieve the system which takes into account the effective scope of the data. We focus on effective life and area as well as popularity of data, and propose a data replacement algorithm of

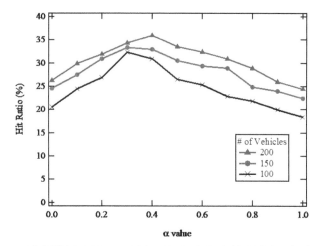

Figure 12. Impact of α value.

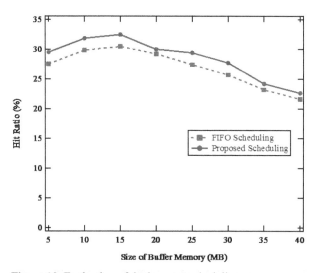

Figure 13. Evaluation of deployment scheduling.

full buffer inside a vehicle and also scheduling method of data deployment from vehicle based on these attributes.

From our simulation results, it is possible for our proposal to distribute more valid data to the effective area within the effective life of data. In the proposed method, data is gradually deleted from vehicles' buffer before its expiration. Our proposal can suppress that the vehicle buffer is occupied by invalid data, and release the useless buffer resources of the vehicle for other more valid data. Thus, our proposal can greatly improve acquisition performance of the time and area specific data.

In the future, it is necessary to set up a proper effective life and area according to the kind of content.

References

[1]   Figueiredo, L., Jesus, I., Tenreiro Machado, J.A., Rui Ferreira, J. and Martins de Carvalho, J. L. (2001) Towards the Development of Intelligent Transport Systems. *Proceedings of IEEE Intelligent Transport Systems Conference*, Oakland, 25-29 August 2001, 1206-1211.

[2]   Zhao, J., Zhang, Y. and Cao, G. (2007) Data Pouring and Buffering on the Road: A New Data Dissemination Paradigm for Vehicular Ad Hoc Networks. *IEEE Transactions on Vehicular Technology*, **56**, 3266-3277.

[3]   Wu, H., Fujimoto, R., Guensler, R. and Hunter, M. (2004) Mddv: A Mobility-Centric Data Dissemination Algorithm for Vehicular Networks. *Proceedings of the 1st ACM International Workshop on Vehicular Ad Hoc Networks*, Philadelphia, 1 October 2004, 47-56.

[4]   Guo, M., Ammar, M. and Zegura, E. (2005) V3: A Vehicle-to-Vehicle Live Video Streaming Architecture. *3rd IEEE International Conference on Pervasive Computing and Communications*, Kauai, 8-12 March 2005, 171-180.

[5]   (2003) Standard Specification for Telecommunications and Information Exchange between Roadside and Vehicle Systems—5GHz Band Dedicated Short Range Communications (DSRC) Medium Access Control (MAC) and Physical Layer (PHY) Specifications. ASTM E2213-03.

[6]   Bluetooth SIG. http://www.bluetooth.org/

[7]   Zhao, J. and Cao, G. (2006) VADD: Vehicle-Assisted Data Delivery in Vehicular Ad Hoc Networks. *IEEE Transactions on Vehicular Technology*, **57**, 1-12.

[8]   Wisitpongphan, N., Bai, F., Mudalige, P. and Sadekar, V. (2007) Routing in Sparse Vehicular Ad Hoc Wireless Networks. *IEEE Journal on Selected Areas in Communications*, **25**, 1538-1566.

[9]   Zhang, Z. (2006) Routing in Intermittently Connected Mobile Ad Hoc Networks and Delay Tolerant Networks: Overview and Challenges. *IEEE Communications Surveys & Tutorials*, **8**, 24-37.

[10] Burgess, J., Gallagher, B., Jensen, D. and Levine, B.N. (2006) Maxprop: Routing for Vehicle-Based Disruption-Tolerant Networks. *Proceedings of the* 25*th IEEE International Conference on Computer Communications*, Barcelona, 23-29 April 2006, 1-11.

[11] Lee, U., Park, J.-S., Yeh, J., Pau, G. and Gerla, M. (2006) Code Torrent: Content Distribution Using Network Coding in VANET. *Proceedings of the* 1*st International Workshop on Decentralized Resource Sharing in Mobile Computing and Networking*, Los Angeles, 25 September 2006, 1-5.

[12] Fiore, M., Casetti, C. and Chiasserini, C.-F. (2005) On-Demand Content Delivery in Vehicular Wireless Networks. *Proceedings of the 8th ACM International Symposium on Modeling, Analysis and Simulation of Wireless and Mobile Systems*, Montréal, 10-13 October 2005, 87-94.

[13] Dietzel, S., Petit, J., Kargl, F. and Scheuermann, B. (2014) In-Network Aggregation for Vehicular Ad Hoc Networks. *IEEE Communications Surveys & Tutorials*, **16**, 1909-1932.

[14] Zhang, Y., Zhao, J. and Cao, G. (2007) On Scheduling Vehicle Roadside Data Access. *Proceedings of the* 4*th ACM International Workshop on Vehicular Ad Hoc Networks*, Montréal, 10 September 2007, 9-18.

[15] Zhang, Y., Zhao, J. and Cao, G. (2009) Roadcast: A Popularity Aware Content Sharing Scheme in Vanets. 29*th IEEE International Conference on Distributed Computing Systems*, Montreal, 22-26 June 2009, 223-230.

[16] Cao, P. and Irani, S. (1997) Cost-Aware WWW Proxy Caching Algorithms. *Proceedings of the USENIX Symposium on Internet Technologies and Systems*, Monterey, 8-11 December 1997, 193-206.

[17] ONE Simulator. http://www.netlab.tkk.fi/tutkimus/dtn/theone/

[18] Breslau, L., *et al.* (1999) Web Caching and Zipf-Like Distributions: Evidence and Implication. *Proceedings of IEEE* 18*th Annual Joint Conference of the IEEE Computer and Communications Societies*, New York, 21-25 March 1999, 126-134.

14

Effect of Transmission Control Protocol on Limited Buffer Cognitive Radio Relay Node

Mohsen M. Tantawy

National Telecommunication Institute (NTI), Cairo, Egypt
Email: ntimohsen@gmail.com

Abstract

Transmission Control Protocol (TCP) is the most important transport layer protocol being used nowadays. It suffers from many problems over mobile networks especially over Cognitive Radio (CR). CR is one of the latest mobile technologies that brings its own share of problems for TCP. The buffer overflow for CR secondary network relay node can affect the performance of TCP. The contribution of this paper is the novel cross-layer model being used to evaluate the effect of the TCP congestion control on the secondary relay node buffer size in Cognitive Radio Network (CRN). The performance has been assessed by buffer overflow probability.

Keywords

TCP, Congestion Control, Cognitive Radio Networks, Cross-Layer Design, Buffer Overflow

1. Introduction

Transmission control protocol (TCP) [1] is the most famous transport protocols that provides a connection which is oriented and reliable end-to-end services with the help of its flow and congestion control mechanisms. However, some of CRNs unique features make TCP performance degrade. In TCP, the receiver replies to the sender with acknowledge that the segment has been received correctly. More than one ACK identifying the same segment to be retransmitted is called a duplicate ACK. After three duplicate ACKs, the sender assumes that the segment has been lost and retransmitted it. TCP also uses timeout to detect losses. After transmitting a segment, TCP starts a time down counter to monitor timeout occurrence. If timeout occurs before receiving the ACK, then the sender assumes that the segment has been lost. The timeout interval is called retransmission timeout (RTO) and is computed according to [2]. This RTO is computed depending on the estimation of Round Trip Time (RTT). TCP congestion window increase is interrupted when a loss is detected.

Two mechanisms are available for the detection of losses: the expiration of an RTO or the receipt of three duplicate ACKs. The source supposes that the network is in congestion and sets its estimate of the capacity to half the current window and the congestion window is reduced to one maximum segment size and the slow start mechanism starts again. When the sender receives three duplicate ACKs, these interact with fast retransmissions and fast recovery mechanisms. TCP Tahoe congestion control [3] sets the window to one packet and uses slow start to arrive the new threshold. When the loss is detected via timeout, ACKs still arrive at the source and losses can be recovered without slow start by using several TCP congestion algorithms like Reno [4], New Reno [5], SACK [6], and Vegas [7] that call a fast recovery algorithm. Once losses are recovered, this algorithm ends and normal congestion avoidance algorithm is called.

Some of these CRNs unique features are primary users (PUs) arrivals, spectrum sensing, spectrum changing, and heterogeneous available channels in secondary users (SUs). It is well-known that TCP has a degraded throughput under wireless systems especially with a high packet loss rate [8]. Due to the existence of PUs in CRNs, a new interruption loss is appeared that causes segment loss and timeout that can cause congestion problems for TCP [9].

Using relay node in cognitive radio to extended CRNs coverage increases the possibility of session initiation between SUs. This relay node in some CRNs scenarios can switch between two different rates due to several parameters, one of them is changing the operation mode from opportunistic to relay on causing under certain conditions relay node buffer overflow. This paper studies the cross-layers interaction between TCP behaviour and relay node buffer overflow.

This paper is organized as follows. Section 2 presents the related works. Section 3 introduces the system models and assumptions. Section 4 gives the numerical results and analysis, and Section 5 concludes the work.

2. Related Works

Authors in [10]-[13] proposed protocols for improving TCP performance over CRNs. The authors in [10] improved TCP throughput compared to other approaches that increase the physical layer throughput. In [11], the authors showed that TCP throughput can be substantially improved if the low-layer parameters in CRN are optimized jointly. The authors in [12] proposed a window-based transport protocol for CR ad-hoc networks that combine the feedback from the intermediate nodes and the destination. Sarkar *et al.* [13] proposed a protocol for the transport layer of the CRN that serve delay-tolerant applications. In [14], the authors create a cross-layer solution considering the effect of bandwidth variation on TCP performance in single hop CRNs without considering spectrum switching. The authors in [9] presents a new type of loss called TCP service interruption with New Reno over CRNs but the authors did not consider the cross layering architecture. In [15], the authors use a cross-layer scheme to improve the energy efficiency of TCP traffic considering the lower layers' characteristics without considering TCP mechanisms or cross-layering. Also in [16] the authors modify the BS that connects TCP over the Internet to a CRN by Local loss recovery and split TCP connection without considering TCP congestion control algorithms. Although, the authors in [13] use a cross-layer approach to serve delay-tolerant applications and to adjust the congestion window by considering spectrum sensing and bandwidth variations, they did not consider the buffer overflow probability. In [17] and [18] authors provide an enhancement for TCP throughput performance considering PUs' activities and lower-layer configurations without considering loss recovery.

3. System Model

In **Figure 1**, we consider a CRN with the coexistence of primary and secondary networks. In the primary network, a primary transmitter sends data to a primary receiver. Meanwhile, in the secondary network, a secondary source communicates with a secondary destination assisted by a finite buffer relay node. The relay node switches the bit rate between the opportunistic rate r_1 and the underlay rate r_2. The relay buffer occupancy B_p can be extracted from the CRN Round Trip Time (*RTT*) as following

$$B_p = (r_1 - r_2) RTT \tag{1}$$

But the maximum buffer occupancy $B_{\max}$ given by

$$B_{\max} = B_c + B_p \tag{2}$$

where B_c is the buffer occupancy at time of rate changing from r_1 to r_2 [19].

According to the model shown in **Figure 1**:

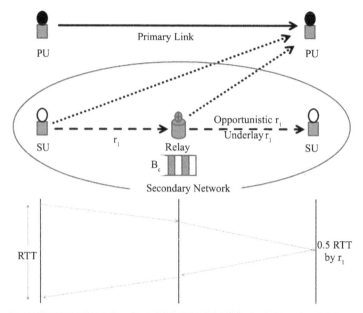

Figure 1. The underlay relay-assisted CRN with delay diagram.

$$W_c = 0.5RTTr_1 + B_c \tag{3}$$

where W_c is the congestion window size, and according to (1) and (3), B_{max} and W_c can be given by

$$B_{max} = W_c + RTT(0.5r_1 - r_2) \tag{4}$$

$$W_c = B_{max} - RTT(0.5r_1 - r_2) \tag{5}$$

The overall probability for the cognitive radio secondary network relay buffer is given by

$$P_B = \frac{\text{the period of time that cause buffer overflow of size } B_p}{\text{the buffer growth cycle period}} = \frac{RTT}{T_p} \tag{6}$$

By using Reno in TCP congestion control with congestion avoidance phase as in **Figure 2**, and at the state of overflow, the window size increased from critical window size W_c to peak window size W_p with changing in time from T_c to $T_c + RTT$ and changing in rate from r_1 to r_2, where $r_1 > r_2$.

$$\int_{W_c}^{W_p} WdW = \int_{T_c}^{T_c+RTT} (r_1 - r_2)dt \tag{7}$$

So, the peak window size can be given by

$$W_p = \sqrt{W_c^2 + 2(r_1 - r_2)RTT} \tag{8}$$

During TCP congestion avoidance phase, the line slop of the congestion avoidance phase is constant so,

$$\frac{W_{max} - W_c}{T_p} = \frac{W_p - W_c}{RTT} \tag{9}$$

From (5), (6), and (9), P_B can be given by

$$P_B = \frac{RTT}{T_p} = \frac{W_p - W_c}{W_{max} - W_c} = \frac{\sqrt{W_c^2 + 2RTT(r_1 - r_2)} + RTT(0.5r_1 - r_2) - B_{max}}{W_{max} - B_{max} + RTT(r_1 - r_2)} \tag{10}$$

Assuming that each TCP segment in the transport layer is segmented into N_{fr} frames in the link layer and for simplicity let all data link frames have the same length L_{fr} and each frame required an acknowledge of length L_{ack} and data rate of r. In this case and from the network model given in **Figure 2**, RTT can be calculated as follows

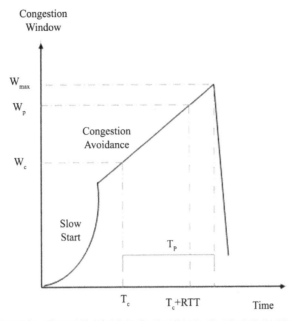

Figure 2. TCP Congestion window with time variation.

$$RTT = N_{fr}\left(\frac{L_{fr} + L_{ack}}{r}\right) \tag{11}$$

Assuming Automatic Repeat Request (ARQ) is used for retransmission, and the average number of retransmissions for each data link frame is N_{re}^{av}, so the round trip time given in (11) can be modified to be

$$RTT = N_{fr}\left(\frac{L_{fr}(N_{re}^{av}+1) + L_{ack}}{r}\right) \tag{12}$$

From [20], the average number of retransmissions for each data link frame N_{re}^{av} can be given by

$$N_{re}^{av} = \frac{F_e - F_e^{N_{re}+1}}{1 - F_e} - N_{re}F_e^{N_{re}+1} \tag{13}$$

where, F_e is the frame error probability given by the bit error rate BER as following

$$F_e = 1 - (1 - BER)^{L_{fr}} \tag{14}$$

4. Numerical Results

In this section, numerical results are presented to evaluate the effect of TCP parameters on buffer overflow. **Table 1** summarizes the numerical values used in the model.

Figure 3 shows the BER versus relay node buffer overflow probability the figure is calculated numerically. Value of CRN opportunistic bit rate is 1 Mbps, and underlay data rate is 0.25 Mbps with 125 K byte maximum TCP window size and 3 frames per TCP segment, ARQ retransmission time is 32. The effect of number of frames per TCP segment is shown in **Figure 4**. The figure shows that increasing TCP segment size increases the number of frames per segment that increase the buffer overflow probability with different BER from less than 10^{-4} to more than 10^{-1}.

Figure 5 shows the maximum buffer size versus relay buffer overflow probability with different BER. The figure shows that value of 5 Kbyte buffer size decrease the overflow probability dramatically from 10^{-2} to 10^{-5}. **Figure 6** shows the effect of maximum congestion window size on buffer overflow with different BER and 500 byte frame and ACK size with and ACK segment with 1 Mbps and 0.25 Mbps opportunistic and underlay modes data rates and 3 frames per segment.

Table 1. System Parameters.

System Parameters	Value
BER	10^{-6} to 10^{-3}
W_{max}	Up to 250 KByte
CR relay node opportunistic mode data rate	1 Mbps
CR relay node underlay mode data rate	0.25 Mbps
Maximum number of ARQ retransmissions	32
Number of frames per TCP segment	3, 5
Frame size	500 Byte
ACK size	40, 500 Byte

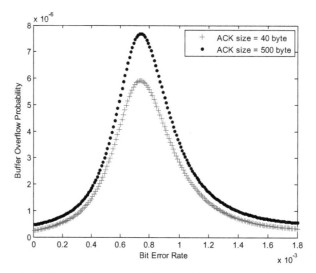

Figure 3. BER vs. Buffer Overflow with variable ACK size.

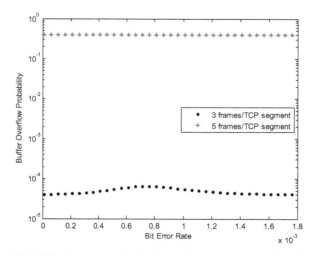

Figure 4. BER vs. Buffer Overflow with variable frames per TCP segment.

5. Conclusion

Most existing CRNs have assumed that there are no effects for TCP behavior on relay node buffer overflow. In

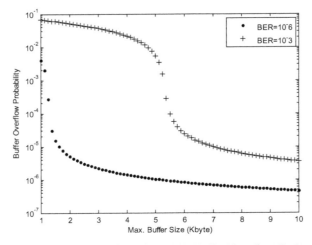

Figure 5. Maximum Buffer size vs. Buffer Overflow Probability.

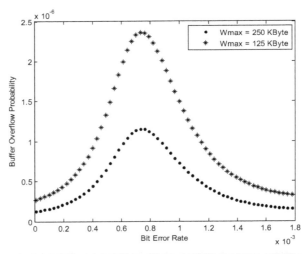

Figure 6. Effect of TCP maximum window size on Buffer Overflow Probability.

this paper, a novel cross-layer model is used to evaluate the effect of the TCP congestion control on the secondary relay node buffer size in Cognitive CRN. The results show that increasing the buffer size decreases the buffer overflow probability for the CR secondary network relay node with value of 5 Kbyte buffer size, and the probability of overflow decreases dramatically from 10^{-2} to 10^{-5}. Moreover, and as expected, increasing BER increases overflow probability. Increasing the number of frames per TCP segment from 3 to 5 frame increases the buffer overflow probability from less than 0.005% up to 50%.

References

[1] Postel, J. (1981) Transmission Control Protocol. RFC 793, Internet Engineering Task Force (IETF).

[2] Paxson, V. and Allman, M. (2000) Computing TCP's Retransmission Timer. RFC 2988, Internet Engineering Task Force (IETF).

[3] Kim, B., Kim, Y.-H., Oh, M.-S. and Choi, J.-S. (2005) Microscopic Behaviors of TCP Loss Recovery Using Lost Retransmission Detection. *2nd IEEE Consumer Communications and Networking Conference (CCNC)*, South Korea, 3-6 January 2005, 296-301. http://dx.doi.org/10.1109/CCNC.2005.1405186

[4] Jacobson, V. (1988) Congestion Avoidance and Control. *Computer Communication Review*, **18**, 314-329. http://dx.doi.org/10.1145/52325.52356

[5] Floyd, S. and Henderson, B. (1999) The New Reno Modifications to TCP's Fast Recovery Algorithm. RFC 2582, In-

ternet Engineering Task Force (IETF).

[6]   Matins, M., Floyd, S. and Romanow, A. (1996) TCP Selective Acknowledgment Options. RFC 2018, Internet Engineering Task Force (IETF).

[7]   Brakmo, L.S. and Paterson, L.L. (1995) TCP Vegas: End to End Congestion Avoidance on a Global Internet. *IEEE Journal on Selected Areas in Communications*, **13**, 1465-1480. http://dx.doi.org/10.1109/49.464716

[8]   Zhong, X.X., Qin, Y. and Li, L. (2014) Transport Protocols in Cognitive Radio Networks: A Survey. *KSII Transactions on Internet and Information Systems*, **8**, 3711-3730.

[9]   Issariyakul, T., Pillutla, L.S. and Krishnamurthy, V. (2009) Tuning Radio Resource in an Overlay Cognitive Radio Network for TCP: Greed Isn't Good. *IEEE Communications Magazine*, **47**, 57-63. http://dx.doi.org/10.1109/MCOM.2009.5183473

[10] Luo, C.Q., Richard, F., Yu, H.J. and Leung, V.C.M. (2009) Optimal Channel Access for TCP Performance Improvement in Cognitive Radio Networks: A Cross-Layer Design Approach. *IEEE Conference on Global Telecommunications*, Honolulu, 30 November-4 December 2009, 2618-2623.

[11] Luo, C.Q., Richard, F., Yu, H.J. and Leung, V.C.M. (2009) Cross-Layer Design for TCP Performance Improvement in Cognitive Radio Networks. *IEEE Vehicular Technology*, **59**, 2485-2495.

[12] Chowdhury, K.R., Di Felice, M. and Akyildiz, I.F. (2009) TP-CRAHN: A Transport Protocol for Cognitive Radio Ad-Hoc Networks. *Proceedings of IEEE Conference on Computer Communications*, Rio de Janeiro, 19-25 April 2009, 2482-2490. http://dx.doi.org/10.1109/INFCOM.2009.5062176

[13] Sarkar, D. and Naray, H. (2010) Transport Layer Protocols for Cognitive Networks. *INFOCOM IEEE Conference on Computer Communications Workshops*, San Diego, 15-19 March 2010, 1-6.

[14] Cheng, Y.C., Wu, E.H. and Chen, G.H. (2010) A New Wireless TCP Issue in Cognitive Radio Networks. *1st International Conference on Networking and Computing*, Higashi-Hiroshima, 17-19 November 2010, 49-54. http://dx.doi.org/10.1109/IC-NC.2010.37

[15] Li, G.Y., Hu, Z., Zhang, G.Y., Zhao, L.L., Li, W. and Tian, H. (2011) Cross-Layer Design for Energy Efficiency of TCP Traffic in Cognitive Radio Networks. *IEEE Vehicular Technology Conference* (*VTC Fall*), San Francisco, 5-8 September 2011, 1-5.

[16] Amjad, M.F., Aslam, B. and Zou. C. (2013) Transparent Cross-Layer Solutions for Throughput Boost in Cognitive Radio Networks. 2013 *IEEE Consumer Communications and Networking Conference* (*CCNC*), Las Vegas, 11-14 January 2013, 580-586.

[17] Wang, J., Huang, A.P,, Wang, W. and Zhang, Z.Y. (2011) Analysis of TCP Throughput in Cognitive Radio Networks. 2011 *IEEE GLOBECOM Workshops* (*GC Wkshps*), Houston, 5-9 December 2011, 930-935. http://dx.doi.org/10.1109/GLOCOMW.2011.6162593

[18] Wang, J., Huang, A.P. and Wang, W. (2012) TCP Throughput Enhancement for Cognitive Radio Networks through Lower-Layer Configurations. *Proceedings of the WSEAS International Conference on Communications*, **48**, 1424-1429. http://dx.doi.org/10.1109/JQE.2012.2217315

[19] Fayed, I.M., Tantawy, M.M., Elbadawy, H.M. and Elramly, S. (2010) Retransmission Time out Probability for TCP-Based Applications over *HSDPA+*. *Proceedings of the World Scientific and Engineering Academy and Society*, 14*th WSEAS International Conference on communications*, Greece, 23-25 July 2010, 189-196.

[20] Chen, D., Ji, H. and Leung, V.C.M. (2012) Distributed Best-Relay Selection for Improving TCP Performance over Cognitive Radio Networks: A Cross-Layer Design Approach. *IEEE Journal on Selected Areas in communication*, **30**, 315-322. http://dx.doi.org/10.1109/JSAC.2012.120210

15

The Impact of Social Media Networks Websites Usage on Students' Academic Performance

Mahmoud Maqableh[1], Lama Rajab[2], Walaa Quteshat[2], Ra'ed Moh'd Taisir Masa'deh[1], Tahani Khatib[2], Huda Karajeh[2]

[1]Management Information Systems, Faculty of Business, The University of Jordan, Amman, Jordan
[2]Computer Information Systems, King Abdullah II School for Information Technology, University of Jordan, Amman, Jordan
Email: maqableh@ju.edu.jo

Abstract

Social Networks Sites (SNSs) are dominating all internet users' generations, especially the students' communities. Consequently, academic institutions are increasingly using SNSs which leads to emerge a crucial question regarding the impact of SNSs on students' academic performance. This research investigates how and to what degree the use of SNSs affects the students' academic performance. The current research's data was conducted by using drop and collect surveys on a large population from the University of Jordan. 366 undergraduate students answered the survey from different faculties at the university. In order to study the impact of SNSs on student's academic performance, the research hypotheses was tested by using descriptive analysis, T-test and ANOVA. Research results showed that there was a significant impact of SNS on the student's academic performance. Also, there was a significant impact of SNS use per week on the student's academic performance, whereas no differences found in the impact of use of SNSs on academic performance due to age, academic achievement, and use per day to most used sites. The findings of this research can be used to suggest future strategies in enhancing student's awareness in efficient time management and better multitasking that can lead to improving study activities and academic achievements.

Keywords

Social Network Sites, Academic Performance, University Students, T-Test, ANOVA

1. Introduction

Social Network Sites (SNSs) have attracted millions of Internet users, who have integrated these sites in their daily lives routines. Twitter and Facebook are among the most popular social networks where the students spend most of their times [1]-[3]. LinkedIn is an example of Social Network site that is used by many students, instructors and scholars for academic purposes. Social Media Network sites can have a positive or negative impact on students' academic performance. However, time management is the factor that contributed towards negative academic performance besides excessive social media use [1].

According to the US Digital Year report, individuals' overall time spent online was among four different online contents including portals, social networking, web-based email's market, and entertainment category (TV and music content, video sites and entertainment news); the second largest share of time spent was on SNS with 14.4% of time spent [4]. University students made up the major proportion of the online networking community. In addition, given the popularity of SNSs, many professors are beginning to use SNSs for enhancing communication with and among students in their classes, class discussions, and teamwork on projects to improve learning outcomes. However, many recent researches pointed out that students' addiction on SNSs can negatively affect student academic performance [1] [5] [6].

Many researchers are studying the effect of rapid and heavy communication technology used by students on their academic performance [5]-[10]. Although many research results pointed to a negative impact of SNSs usage on academic performance [1] [5] [6]; yet, some researchers found a little or no negative effects on students' academic performance if good multitasking is achieved by students [6]. Therefore, students could have better control on SNSs use if they have more self-regulations and enhanced time management skills [11]. A research group suggested that online social networks could possibly be viewed as a helpful educational technology if more academic staff actually knew how to incorporate them into their curricula [5]. They presented a research model that defines the direct and indirect key factors that affect academic performance of students, particularly, the impact of time spent on social media. The model indicates that students' academic performance is a function of attention span, time management skills, student characteristics, academic competence and time spent on online social networks.

This research aims to add a better clarity to this research area by examining the relationship between the use of SNSs and students' academic performance. Hence, the main objective of our study is to investigate whether there is a positive impact of the use of social network website on undergraduate students' academic performance from all faculties in the University of Jordan. The structure of this paper is as follows. Section 2 reviews the literature and related work to the research. Section 3 describes the research methodology. Section 4 presents the data analysis and results. Section 5 gives further discussion of the findings and concludes the paper.

2. Literature Review

Internet is being used by an increasing portion of world's population on a regular and daily basis. Technology is being implemented in most fields of our lives such as education, entertainment, and commerce [12] [13]. Many researchers studied the significance of using social network sites (SNSs), and its impact on the academic performance of high school or college students. Some of these studies focused on the effects of factors such as multitasking, time management, student characteristics and personality, study system and strategy, and academic competency [5]-[10]. Moreover, some researchers studied the cultural difference of SNS patterns uses, the attitudes of users toward SNSs, and their perspective of SNS use [14]. In the following subsections we will discuss the significance of SNSs to undergraduate student, the impact of SNSs on academic performance, and the role of multitasking.

2.1. The Significance of SNSs to Undergraduate Students

The academic and social importance of SNSs for undergraduate students was pointed out in literature. Michikyan *et al.*, argued that college students need to develop new networks of support; especially for those with new academic experiences such as: first-year students, first generation college students, and immigrant students [3]. These students may use online network sites such as Facebook, Twitter or LinkedIn to develop the support network they need [15]. Furthermore, social media helps in establishing peer-support networks prior to first-year students arriving to campus.

As for first generation students, whom parents have not attended university or undergraduate school, social media is very important because it offers emotional support and confidence from Facebook friends [14]. First generation lack parental emotional support given that their parents have not experienced successfully graduating from college and are less involved in society and with teachers [16]. The students' confidence of college application process, for instance, was associated with instrumental support from Facebook friends and information obtained from social media [14]. Accordingly, SNSs are used frequently by more students to build new friendship networks in order to be successful in both social and academic domains [17]. SNSs also help in facilitating the art of learning by providing a media to share ideas which allow students to collaborate with others through building their own virtual communities [1]. [14] stated that Facebook as the most popular example of SNSs plays a crucial role in predicting expectations of college success; *i.e.* having successful Facebook colleagues in college serve a positive example of people from similar backgrounds.

2.2. The Impact of SNSs on Academic Performance

Although SNSs is a very helpful tool in students' hands, it was found by many studies that a negative impact of social network sites usage on academic performance could occur [1] [5] [6]. Many students claimed that SNSs usage did not affect their marks, while others admit that SNSs permanent usage can be a distraction, time consuming, and lead to academic procrastination [1] [18]. In [5] research, results pointed to a negative impact of online social media usage on academic performance; therefore, as time spent on social networking sites increases, the academic performance of the students is seen to deteriorate. Nevertheless; (Junco, 2015) studied the impact of college students' academic level and found that Facebook affected Grade Point Averages (GPAs) negatively for freshmen, sophomores, juniors but not for seniors. Seniors spend less time on Facebook, and they are less likely to post status updates comments, chats, posts, videos or photos than others [10].

On the other hand; as SNSs can be accessed from digital device such as a personal computer, a smart phone, a tablet or a laptop; some studies focused on the impact of using such digital devices on students' academic performance since those devices are the main reason for the excessive usage of SNSs. [6], for instance, said that there is a strong negative correlation between the amount of time spent on the computer and the time spent on study; and again they found that people who spend more time on computers spend less time on studying and get less GPAs. [19] explored that Cell Phone use was negatively related to GPA and positively related to anxiety. They found that high frequency cell phone users tended to have lower GPA, higher anxiety, and lower satisfaction with life relative to their peers who used the cell phone less often.

2.3. The Role of Multitasking

Although many researchers say that there is a negative impact between Facebook usage and academic performance, some researchers found that the increase in use of social networks did not necessarily decrease the academic performance of the students [1]. They found that most students did not use social media excessively, however there were some exceptions. In fact, the process of managing time between study and social network usage, known also as multitasking, can mitigate the risk of lower GPAs due to Facebook use. Consequently, in order to be successful, students should learn how to balance academic and social demands. In [5], authors even suggested that instructors could use mandatory policies disallowing students to use their phones and computers unless required for course purposes.

Many students think that time management is the factor that contributed towards negative academic performance besides excessive social media use [1]. However, some researchers found that although students feel competent in their ability to use social networks sites for academic purposes, they did not have the desire or willingness to do so [5]. As mentioned before, [10] pointed out that seniors use Facebook less than freshmen and they are better in multitasking, since they have understood what they need to do in order to be successful. Researchers suggested that first year students must learn how to balance academic and social demands effectively in order to be successful. It is important that these new students should do such balancing through building social connections. Generally speaking, students with more self-regulations were more able to control Facebook use [11].

3. Research Methodology

The major elements of this research are established based on preceding literature, either theoretically or empirically. This section provides the methodology applied in this study. The methodology includes the research theoretical framework, procedural definitions, research hypotheses, research type and scale, research population and sample, besides data collection and analysis procedures. The reliability and validity of the study are also provided. As this research is deductive and quantitative in its nature, one of the important characteristic of deduction is the need to operationalize the variables of the study in a way that facilitate the measurement of facts quantitatively [20]. Indeed, the independent variable of the use of social networks websites and the dependent variable of academic achievements were identified from [1] [5] [6] [10].

3.1. Research Theoretical Model

This study used variables that are common in social network literature. By reviewing the literature, it was noticed that there is a need to examine if the use of social networks websites among universities students could impact their academic achievements directly or moderated by other variable. **Figure 1** displays the research's proposed model.

[21] defines social network sites as web-based services that allow individuals to construct a public or semi-public profile within a bounded system, articulate a list of other users with whom they share a connection, and view and traverse their list of connections and those made by others within the system. Indeed, the use of social networks websites may be defined as the degree on which users spend time using social network websites and academic as facts. However, we can define academic achievement as student's performance and to which extent he/she obtained the education objectives, goals and outcomes.

3.2. Research Hypotheses

In order to test the causal model of the relationship between the uses of social networks websites and academic achievements; the study is hypothesized as follows:

3.2.1. The Main Research Hypotheses

H1: There is a statistically significant impact of the use of social networks websites on academic achievements.

H2: There is a significant difference in the impact of the use of social networks websites on academic achievements due to demographic characteristics.

3.2.2. The Research Sub-Hypotheses

H2A: There is a significant difference in the impact of the use of social networks websites on academic achievements due to gender.

H2B: There is a significant difference in the impact of social networks websites on academic achievements due to age.

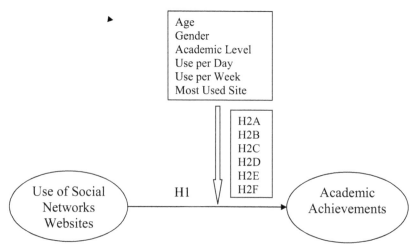

Figure 1. The research model.

H2C: There is a significant difference in the impact of social networks websites on academic achievements due to academic level.

H2D: There is a significant difference in the impact of social networks websites on academic achievements due to use per day.

H2E: There is a significant difference in the impact of social networks websites on academic achievements due to use per week.

H2F: There is a significant difference in the impact of social networks websites on academic achievements due to most used site.

3.3. Population and Sample

Data should be collected from the people that can provide the correct answers to solve the problem [22] and represent the whole people, events or objects the researcher want to study. Therefore, the population of this study consists of bachelor students who studied Social Media Network course as elective course from the University of Jordan located in Jordan, which counts of more than 30,000 students according to the university's registration unit. Indeed, the researchers have chosen bachelor students from all faculties as they are the largest category studying at the University of Jordan. Also, the current research data was conducted by using drop and collect surveys which covers large samples of the population. This technique is less expensive and consumes less time than other methods such as interviews; and covers a wider geographical area. As a result, the researchers used this method of data collection at the University of Jordan. Indeed, by using a drop and collect method to thousands of students, 366 survey questionnaires were returned, which is adequate for statistical analysis, based on [23] [24].

4. Data Analysis and Results

Descriptive analysis was used to describe the characteristic of sample and the respondent to the questionnaires. Correlation coefficients were used to determine the relationship between the items for both dependent and independent variables, where the abbreviations of each of the study's variables are as follow: USS: use of social networks websites; and AAS: academic achievements. In addition, a simple linear regression, T-test and ANOVA test were employed to test the hypotheses.

4.1. Reliability and Validity

According to [25] it is important to make sure that the instrument developed to measure a particular concept is accurately measuring the variable and is actually measuring the concept that it is supposed to measure in the research. Indeed, reliability analysis is related to the assessment of the degree of consistency between multiple measurements of a variable, whereas validity analysis refers to the degree to which a scale or set of measures accurately represents the construct [26]. The reliability of the instrument was measured by the Cronbach's alpha coefficient. Further, some scholars e.g. [27] suggested that the values of all indicators or dimensional scales should be above the recommended value of 0.60. However, the Cronbach's alpha for the independent variable (*i.e.* use of social networks websites) was 0.87; while the dependent variable, academic achievements, has a Cronbach's alpha coefficient of 0.86.

Convergent validity refers to the degree to which items or measures are correlated with each other to measure the same construct. Therefore, higher correlation shows that the scale is assessing its aimed construct. The closer the values are to 1 the more highly correlated the items are, and specifically the individual item reliability is recommended to be greater than 0.50 [27] [28]. **Table 1** and **Table 2** display the correlations between the independent items, and the dependent items respectively.

It has been noticed from both **Table 1** and **Table 2** that most of the values were close to 0.5 which indicate a positive correlation between items. This leads to a higher level of convergent validity.

4.2. Respondents Demographic Profile

As showed in **Table 3**, the demographic profile of the respondents for this study revealed that they are typically female, most of them in their second and third year level, about 64% of them are of ages between 20-less than 23 years old, and using Facebook heavily more over an hour daily and more than 10 hours per week.

Table 1. Correlations between the independent items.

USS Items	USS1	USS2	USS3	USS4	USS5	USS6	USS7
USS1	1.000						
USS2	0.577	1.000					
USS3	0.702	0.597	1.000				
USS4	0.483	0.340	0.475	1.000			
USS5	0.682	0.480	0.643	0.445	1.000		
USS6	0.530	0.390	0.530	0.435	0.587	1.000	

Table 2. Correlations between the dependent items.

AAS Items	AAS1	AAS2	AAS3	AAS4	AAS5	AAS6	AAS7
AAS1	1.000						
AAS2	0.354	1.000					
AAS3	0.408	0.568	1.000				
AAS4	0.432	0.433	0.566	1.000			
AAS5	0.448	0.476	0.577	0.588	1.000		
AAS6	0.567	0.365	0.462	0.546	0.482	1.000	
AAS7	0.406	0.356	0.498	0.508	0.439	0.493	1.000

Table 3. Respondents demographic profile based on the researchers' analysis.

Category Name of the Company	Frequency	Percentage %
Gender		
Male	107	29.2
Female	259	70.9
Total	366	100
Age		
17 years—less than 20	121	33.1
20 years—less than 23	233	63.7
24 years—less than 26	9	2.5
27 years and above	3	0.8
Total	*366*	*100*
Educational Level		
Year 1	27	7.4
Year 2	137	37.4
Year 3	121	33.1
Year 4	75	20.5
Year 5 and more	6	1.6
Total	366	100

Continued

Use Per Day		
Less than half an hour	13	3.6
Half an hour—less an hour	65	17.8
One hour—less than 3 hours	147	40.2
3 hours and more	141	38.5
Total	366	100
Use Per Week		
Less than 10 hours	88	24.0
10–less than 29 hours	148	40.4
29–less than 50 hours	89	24.3
50 hours and more	41	11.2
Total	366	100
Most Used Site		
You Tube	19	5.2
Twitter	6	1.6
Facebook	221	60.4
WhatsApp	111	30.3
Others	9	2.5
Total	366	100

4.3. Descriptive Analysis

In order to describe the responses and thus the attitude of the respondents toward each question they were asked in the survey, the mean and the standard deviation were estimated. While the mean shows the central tendency of the data, the standard deviation measures the dispersion which offers an index of the spread or variability in the data [22] [25]. In other words, a small standard deviation for a set of values reveals that these values are clustered closely about the mean or located close to it; a large standard deviation indicates the opposite. **Table 4** shows the overall mean and standard deviation of the independent and dependent variables.

As presented in **Table 4**, data analysis results have shown that the use of social networks websites among bachelor students is applied to a great extent at the University of Jordan in which the mean score is 3.7233. This serves as an indicator on the importance of the use of social networks websites and the essential role that they play in enhancing the students' academic achievements. Furthermore, data analysis results have revealed that academic achievements itself is applied to a great extent at the University of Jordan in which the mean score is 3.7459. This high level of presentation denotes a positive attitude regarding students' academic achievements. This sturdily advocates that University of Jordan is currently engaging in information technology and mobile application activities with their students.

4.3.1. Independent Variable
Table 5 demonstrates the mean scores for the use of social networks websites.

4.3.2. Dependent Variable
Table 6 demonstrates the mean scores for academic achievements items.

Table 4. Mean and standard deviation of the study's variables.

Type of Variable	Variables	Mean	Std. Deviation
Independent Variable	Use of social networks websites	3.7233	0.70079
Dependent Variable	Academic achievements	3.7459	0.68193

Table 5. Mean and standard deviation for the use of social networks websites.

Use of social networks websites	Mean	Std. Dev.
Social networking is useful to me as a student.	3.79	0.956
Social networking sites have a positive impact on my Academic Achievement.	3.31	1.108
Social networks help me to achieve my academic goals.	3.65	0.958
The use of social networks help to improve my contact with my colleagues and teachers as well as my performances academic.	3.98	0.848
Skills and knowledge obtained during studying Social Networks Course are very important to my performance and academic achievement.	3.84	0.918
I know the most important concepts and facts relating to social net\works communications have improved.	3.83	0.863
The study of topics related to social networking has a positive impact on my life in the future.	3.68	0.850

Table 6. Mean and standard deviation for academic achievements.

Use of social networks websites	Mean	Std. Dev.
I deal with social networking sites are faster.	3.97	0.980
Currently I use social networking sites more than before and frequently.	3.76	0.935
My performance has improved in the use of social networking sites better than before.	3.72	0.877
Recently, I was introduced to the new features of the social networking sites.	3.74	0.893
Recently, I have developed my use of various social networking sites.	3.62	0.897
Recently, I have developed my skills and knowledge in the use of social networking sites.	3.80	0.891
Recently, I learned how to think and analysis of topics on the subject of social networking critically.	3.62	0.977

4.3.3. The Moderating Variables

The demographic characteristics of student are the moderating variables in this study. These variables are used in order to identify if there are different patterns between academic achievements among respondents. The demographic characteristics in this study include gender, age, academic level, use per day, use per week, and most used site.

4.4. Hypothesis Testing Results

The purpose of this study was to test the impact of the use of social networks websites on academic achievements. Thus, in order to test the hypotheses developed for this study, a simple linear regression technique was used. Further, the level of significance (α-level) was chosen to be 0.05 and the probability value (p-value) obtained from the statistical hypotheses test is considered to be the decision rule for rejecting the null hypotheses [29]. If the p-value is less than or equal to α-level, the null hypothesis will be rejected and the alternative hypothesis will be supported. However, if the p-value is greater than the α-level, the null hypothesis cannot be rejected and the alternative hypothesis will not be supported.

4.4.1. Hypothesis 1

H1: There is a statistically significant impact of the use of social networks websites on academic achievements. The results of testing of the main hypothesis are demonstrated in **Tables 7-9**.

Table 7. Study model summary.

Model	R	R Square	Adjusted R Square	Std. Error of the Estimate
1	0.839 a	0.703	0.702	0.37196

a: Predictors: (Constant), USS.

Table 8. Analysis of variance for the study model (b).

Model	Sum of Squares	Df	Mean Square	F	Sig.	Result
Regression	119.376	1	119.376	862.852	0.000 a	
Residual	50.360	364	0.138			Accept the hypothesis
Total	169.736	365				

a: Predictors: (Constant), UUS; b: Dependent Variable: AAS.

Table 9. Coefficient of predictors (a).

Model	Unstandardized Coefficients		T	Sig.	Result of Hypothesis Testing
	B	Std. Error			
(Constant)	0.707	0.105	6.722	0.000	
USS	0.816	0.028	29.374	0.000	Accept the hypothesis

Dependent variable: AAS.

The multiple correlation coefficient R = 0.839 shows that there is a positive correlation between the use of social networks websites on academic achievements. This means that the independent variable and the dependent variable change in the same direction. The multiple correlation coefficient is a gauge of how well the model predicts the observed data. The value of R^2 = 0.703 indicates the amount of variations in academic achievements that is accounted by the fitted model. This is to say that 70.3% of the variability of academic achievements has been explained by the variable of the use of social networks websites. Also, the higher the use of social networks websites, the higher the applicability of academic achievements itself. In order to generalize the results obtained from the respondents to the whole population, adjusted R^2 was calculated. Indeed, adjusted R^2 was equals 70.2%, indicating a high degree of generalizability of the model. **Table 8** showed the Analysis of variance (ANOVA) analysis to test the main null hypothesis. Indeed, F-ratio for the data was 862.852 which is significant at $p < 0.05$ (sig = 0.000). Therefore, there was a statistically significant impact of the use of social networks websites on academic achievements, and thus reject the null hypothesis and accept the alternative hypothesis. Moreover, the equation of the single linear regression takes this formula:

$$Y = \beta_0 + \beta_1 X_1 + \varepsilon i \tag{1}$$

By testing the impact of the predictor included in the model (*i.e.* using the value of β and α significance level) on the dependent variable, we can infer the acceptability of the hypothesis. The β indicates the individual contribution of the predictor to the model if the predictor is held constant. **Table 9** shows that for the use of social networks websites; the value of β was 0.816; and considered to be high. In addition, t-value was above 1.96. Indeed, we can infer from the values of β and t-value that the use of social networks websites did have a statistically significant impact on students' academic achievements at the University of Jordan.

4.4.2. Hypothesis 2
Hypotheses H2A, H2B, H2C, H2D, H2E and H2F argued that there is a significant difference in the impact of the use of social networks websites on academic achievements due to gender, age, academic level, use per day, use per week, and most used site respectively. Independent Samples T-test was employed in order to investigate if there any significant differences in the impact of the use of social networks websites on academic achievements that can be attributed to gender. Also, ANOVA test was employed to examine if there any significant differences in the impact the use of social networks websites on academic achievements that can be attributed to age, academic level, use per day, use per week, and most used site.

Results of T-test, shown in **Table 10**, indicated that there is no significant difference in the impact of the use of social networks websites on academic achievements that can be attributed to gender. On the other hand, results of ANOVA test, shown in **Table 11**, **Table 12**, **Table 15** and **Table 16**, indicated that there is no significant difference in the impact of the use of social networks websites on academic achievements in favor of age, academic level, use per week, and most used site; whereas **Table 13** indicated that there is a significant difference in the impact of the use of social networks websites on academic achievements in favor of use per day. Therefore, **Table 14** provided the statistical significance of the differences between each pair of groups for use per day. As noticed in **Table 14**, the four groups (*i.e.* less than half an hour, Half an hour—less an hour, One hour—less than 3 hours, 3 hours and more) were statistically different from one another.

5. Discussion and Conclusions

The primary motivation of this research is to test the relationship between the use of social networks websites and the student's academic performance. The population of this study consists of undergraduates students from all faculties of the University of Jordan. After analyzing the respondents demographic profile, it revealed that they are typically females and most of them in the second and third year level (sophomores and juniors), with ages between 20 - 23 years old. Results also show that they are using Facebook heavily given that 38.5% of students are spending three hours daily, and about 40% of them are spending more than 10 hours per week.

The proposed model has two main hypotheses (H1 and H2). The fist hypothesis claims that there is a direct and significant impact of the use of social networks and the student's academic performance. On the other hand, the second hypothesis with its sub hypotheses (H2A, H2B, H2C, H2D, H2E, and H2F) claims that there is a difference in the impact of use of social networks websites due to demographic characteristics (moderate variables: age, gender, academic level, use per day, use per week, and the most used site). Before testing the proposed hy-

Table 10. T-test of the use of social networks websites on academic achievements attributed to gender.

Variables	Male			Female			T	df	Sig.
	N	Mean	Std. Dev.	N	Mean	Std. Dev.			
Academic achievements	107	3.6182	0.88972	259	3.7987	0.56842	1.911	143.090	0.066

Table 11. ANOVA Analysis of the use of social networks websites on academic achievements attributed to age.

Variables		Sum of Squares	Df	Mean Square	F	Sig.
	Between Groups	0.968	3	0.323	0.692	0.557
Academic achievements	Within Groups	168.768	362	0.466		
	Total	169.736	365			

Table 12. ANOVA Analysis of the use of social networks websites on academic achievements attributed to academic level.

Variables		Sum of Squares	Df	Mean Square	F	Sig.
	Between Groups	0.894	4	0.223	0.478	0.752
Academic achievements	Within Groups	168.842	361	0.468		
	Total	169.736	365			

Table 13. ANOVA Analysis of the use of social networks websites on academic achievements attributed to use per day.

Variables		Sum of Squares	Df	Mean Square	F	Sig.
	Between Groups	9.485	3	3.162	7.142	0.000
Academic achievements	Within Groups	160.252	362	0.443		
	Total	169.736	365			

Table 14. Multiple comparisons analysis of the use of social networks websites on academic achievements attributed to use per day.

(I) use per day	(J) use per day	Mean Difference (I-J)	Std. Error	Sig.	95% Confidence Interval	
					Lower Bound	Upper Bound
Less than half an hour	Half an hour—less an hour	−0.72308*	0.202150	0.002	−1.2448	−0.2013
	One hour—less than 3 hours	−0.75181*	0.19252	0.0010	−1.2487	−0.2549
	3 hours and more	−0.87608*	0.19285	0.0000	−1.3738	−0.3783
Half an hour—less an hour	Less than half an hour	0.72308*	0.20215	0.002	0.2013	1.2448
	One hour—less than 3 hours	−0.02874	0.09911	0.991	−0.2845	0.22710
	3 hours and more	−0.15300	0.09975	0.418	−0.4105	0.1045
One hour—less than 3 hours	Less than half an hour	0.75181*	0.19252	0.001	0.2549	1.2487
	Half an hour—less an hour	0.02874	0.09911	0.991	−0.2271	0.2845
	3 hours and more	−0.12427	0.07843	0.389	−0.3267	.07820
3 hours and more	Less than half an hour	0.87608*	0.19285	0.000	0.3783	1.3738
	Half an hour—less an hour	0.153000	0.09975	0.418	−0.1045	0.4105
	One hour—less than 3 hours	0.12427	0.07843	0.389	−0.0782	0.3267

*The mean difference is significant at the 0.05 level.

potheses, it was important to measure the consistency between the measured variables and how accurately they represent the proposed construct. For this purpose, the reliability and the convergent validity for both independent variable (use of social networks website) and dependent variable (academic achievement) were approved. Cronbach's alpha coefficient and the correlation between the variables showed satisfactory results.

In order to test the hypotheses, a simple linear regression technique, T-test and ANOVA, were used. The main hypothesis H1 was analyzed. The results shown in **Tables 7-9** indicated that there was a statistically significance impact of the use of social networks websites on academic achievement. Since the correlation showed that the use of social networks and the academic performance variables were highly correlated and changed in the same direction. Hypotheses H2A, H2B, H2C, H2D, H2E, and H2F stated that there were significant differences in the impact of the use of social networks websites on academic achievements due to gender, age, academic level, use per day, use per week and most used site, respectively. T-test was used to test hypotheses H2A which stated that there was no significant difference in the impact of the use of social networks on academic achievement due to gender. The p-value (significance level) was 0.066 which is greater than the α level 0.05 as explained in the results section so this means that we fail to reject the hypothesis as was shown in **Table 10**.

The remaining hypotheses (H2B, H2C, H2D, H2E, and H2F) were tested by using the ANOVA test and the significance level for them all is greater than 0.05 which means that there is no significance difference in the impact of use of social networks websites on academic achievements due to age, academic achievement, use per week and to most used sites as shown in **Tables 10-12**, **Table 15** and **Table 16** in the previous section. But surprisingly the significance level of the use per day indicated that there was a significance difference in the impact of use of social networks on the academic achievement due to use per day. Further tests was made between groups: (less than half an hour, half an hour—less and hour, one hour—less than 3 hours, 3 hours and more) as shown in **Table 14**, the results revealed that the higher the number of hours had more effect on the academic achievement which is reasonable because that number of hours per day is limited and the chance for multitasking is also limited which will finally affect the academic performance.

When comparing the use per day with the use per week results, one can obviously notice that there was no impact of using social networks websites per week on the academic achievement. We think that this result can be justified by saying that the time might be utilized more efficiently during the weekend. Students have more free hours during the weekend day, so multitasking can be achieved more and students will compensate the time they wasted during the week and will focus on efficient studying. We recommend that future researches should

Table 15. ANOVA Analysis of the use of social networks websites on academic achievements attributed to use per week.

Variables		Sum of Squares	df	Mean Square	F	Sig.
	Between Groups	2.511	3	0.837	1.812	0.145
Academic achievements	Within Groups	167.226	362	0.462		
	Total	169.736	365			

Table 16. ANOVA Analysis of the use of social networks websites on academic achievements attributed to most used site.

Variables		Sum of Squares	Df	Mean Square	F	Sig.
	Between Groups	1.762	4	0.441	0.947	0.437
Academic achievements	Within Groups	167.974	361	0.465		
	Total	169.736	365			

consider the negative impact of excessive usage of social networks on students' performance and achievements. Consequently more studies should focus on the importance of faculties' workshops and seminars that can be useful in training student to manage the use of social networks more efficiently.

References

[1]   Karpinski, A.C., Kirschner, P.A. Ozer, I., Mellott, J.A. and Ochwo, P. (2013) An Exploration of Social Networking Site Use, Multitasking, and Academic Performance among United States and European University Students. *Computers in Human Behavior*, **29**, 1182-1192. http://dx.doi.org/10.1016/j.chb.2012.10.011

[2]   Alwagait, E., Shahzad, B. and Alim, S. (2014) Impact of Social Media Usage on Students Academic Performance in Saudi Arabia. *Computers in Human Behavior*, **51**, 1092-1097.

[3]   Michikyan, M., Subrahmanyam, K. and Dennis, J. (2015) Facebook Use and Academic Performance among College Students: A Mixed-Methods Study with a Multi-Ethnic Sample. *Computers in Human Behavior*, **45**, 265-272. http://dx.doi.org/10.1016/j.chb.2014.12.033

[4]   ComScore.com (2012) Internet Website. ComScore. http://www.comscore.com/Insights/Presentations-and-hitepapers/2012/2012-US-Digital-Future-in-Focus

[5]   Paul, J.A., Baker, H.M. and Cochran, J.D. (2012) Effect of Online Social Networking on Student Academic Performance. *Computers in Human Behavior*, **28**, 2117-2127. http://dx.doi.org/10.1016/j.chb.2012.06.016

[6]   Wentworth, D.K. and Middleton, J.H. (2014) Technology Use and Academic Performance. *Computers & Education*, **78**, 306-311. http://dx.doi.org/10.1016/j.compedu.2014.06.012

[7]   Lomi, A., Snijders, T.A.B. Steglich, C.E.G. and Torló, V.J. (2011) Why Are Some More Peer than Others? Evidence from a Longitudinal Study of Social Networks and Individual Academic Performance. *Social Science Research*, **40**, 1506-1520. http://dx.doi.org/10.1016/j.ssresearch.2011.06.010

[8]   Mehmood, S. (2013) The Effects of Social Networking Sites on the Academic Performance of Students in College of Applied Sciences, Nizwa, Oman. *International Journal of Arts and Commerce*, **2**, 111-125.

[9]   Ellore, S.B., Niranjan, S. and Brown, U.J. (2014) The Influence of Internet Usage on Academic Performance and Face-to-Face Communication. *Journal of Psychology and Behavioral Science*, **2**, 163-186.

[10] Junco, R. (2015) Student Class Standing, Facebook Use, and Academic Performance. *Journal of Applied Developmental Psychology*, **36**, 18-29. http://dx.doi.org/10.1016/j.appdev.2014.11.001

[11] Rouis, S., Limayem, M. and Salehi-Sangari, E. (2011) Impact of Facebook Usage on Students' Academic Achievement: Role of Self-Regulation and Trust. *Electronic Journal of Research in Educational Psychology*, **9**, 961-994.

[12] Karajeh, H., Maqableh, M. and Masa'deh, R. (2014) A Review on Stereoscopic 3D: Home Entertainment for the Twenty First Century. Vol. 5, Springer, New York, 1-9.

[13] Maqableh, M. (2012) Analysis and Design Security Primitives Based on Chaotic Systems for E-Commerce. Durham University, Durham.

[14] Rienties, B. and Tempelaar, D. (2013) The role of Cultural Dimensions of International and Dutch Students on Academic and Social Integration and Academic Performance in the Netherlands. *International Journal of Intercultural Relations*, **37**, 188-201. http://dx.doi.org/10.1016/j.ijintrel.2012.11.004

[15] Deandrea, D.C., Ellison, N.B., Larose, R., Steinfield, C. and Fiore, A. (2012) Serious Social Media: On the Use of Social Media for Improving Students' Adjustment to College. *Internet and Higher Education*, **15**, 15-23. http://dx.doi.org/10.1016/j.iheduc.2011.05.009

[16] Wohn, D.Y., Ellison, N.B., Khan, M.L., Fewins-Bliss, R. and Gray, R. (2013) The Role of Social Media in Shaping First-Generation High School Students' College Aspirations: A Social Capital Lens. *Computers & Education*, **63**, 424-436. http://dx.doi.org/10.1016/j.compedu.2013.01.004

[17] Ramírez Ortiz, M.G., Caballero Hoyos, J.R. and Ramírez López, M.G. (2004) The Social Networks of Academic Performance in a Student Context of Poverty in Mexico. *Social Networks*, **26**, 175-188. http://dx.doi.org/10.1016/j.socnet.2004.01.010

[18] Ozer, I., Karpinski, A.C. and Kirschner, P.A. (2013) A Cross-Cultural Qualitative Examination of Social-Networking Sites and Academic Performance. *Procedia—Social and Behavioral Sciences*, **112**, 873-881.

[19] Lepp, A., Barkley, J.E. and Karpinski, A.C. (2014) The Relationship between Cell Phone Use, Academic Performance, Anxiety, and Satisfaction with Life in College Students. *Computers in Human Behavior*, **31**, 343-350. http://dx.doi.org/10.1016/j.chb.2013.10.049

[20] Saunders, M., Lewis, P. and Thornhill, A. (2007) Research Methods for Business Students. 4th Edition, Financial Times Prentice Hall, Edinburgh Gate, Harlow.

[21] Boyd, D.M. and Ellison, N.B. (2007) Social Network Sites: Definition, History, and Scholarship. *Journal of Computer-Mediated Communication*, **13**, 210-230. http://dx.doi.org/10.1111/j.1083-6101.2007.00393.x

[22] Sekaran, U. (2003) Research Methods for Business: A Skill-Building Approach. 4th Edition, John Wiley and Sons, Hoboken.

[23] Krejcie, R.V. and Morgan, D.W. (1970) Determining Sample Size for Research Activities. *Educational and Psychological Measurement*, **30**, 607-610.

[24] Pallant, J. (2005) SPSS Survival Guide—A Step by Step Guide to Data Analysis Using SPSS for Windows. Open University Press, Chicago.

[25] Sekaran, U. and Roger, B. (2013) Research Methods for Business: A Skill-Building Approach. 6th Edition, John Wiley and Sons, Hoboken.

[26] Hair, J., Anderson, R., Tatham, R. and Black, W. (1998) Multivariate Data Analysis. Vol. 259, 5th Edition, Prentice-Hall International Inc., Upper Saddle River.

[27] Bagozzi, R. and Yi, Y. (1988) On the Evaluation of Structural Equation Models. *Journal of the Academy of Marketing Science*, **16**, 74-94. http://dx.doi.org/10.1007/BF02723327

[28] Fornell, C. and Larcker, D.F. (1981) Evaluating Structural Equation Models with Unobservable Variables and Measurement Error. *Journal of Marketing Research*, **18**, 39-50.

[29] John, W.C. (2009) Research Design: Qualitative, Quantitative, and Mixed Methods Approaches. 3th Edition, Sage Publications, Thousand Oaks.

16

Performance Study of Locality and Its Impact on Peer-to-Peer Systems

Mohammed Younus Talha, Rabah W. Aldhaheri, Mohammad H. Awedh

Department of Electrical and Computer Engineering, King Abdulaziz University, Jeddah, Saudi Arabia
Email: ymohammed0003@stu.kau.edu.sa, raldhaheri@kau.edu.sa, mhawedh@kau.edu.sa

Abstract

This paper presents the measurement study of locality-aware Peer-to-Peer solutions on Internet Autonomous System (AS) topology by reducing AS hop count and increase nearby source nodes in P2P applications. We evaluate the performance of topology-aware BT system called TopBT with BitTorrent (BT) by constructing AS graph and measure the hops between nodes to observe the impact of quality of service in P2P applications.

Keywords

Peer-to-Peer, BitTorrent, TopBT, Locality, Autonomous System

1. Introduction

Peer-to-Peer (P2P) is a distributed computing model which aims to share resources whose concept is not completely new. However, P2P systems are natural evolution in decentralized system architecture [1] in which peer is a node that can act as a client and server simultaneously in dynamic environment. Nodes can join or leave the system freely and also exchange resources directly without the help of a third party server. Popular P2P systems generate massive amount of traffic over the internet and it has been reported that 65% - 70% Internet backbone is P2P traffic. Furthermore, it may be estimated that 50% - 65% of download traffic and 75% - 90% of upload traffic is generated by P2P traffic access communities [2]. P2P networks can be classified according to their functionalities into three main classes such as file sharing, video streaming and VoIP. File sharing P2P applications like BitTorrent (BT), TopBT are the most popular among the three classes whereas video streaming classes applications are PPLive, PPStream and Voip application is Skype.

Zatto [3] was introduced as a localized P2P live streaming system and Skype [4] was modified to implement locality in super peers selections. Finally, Top-BT [5] was introduced as localized version of the BitTorrent software which is developed by OHIO State University R&D Dept., that actively discovers its network proximities to its connected peers, this unique feature separates it from Bit torrent.

It also improves the peers transmission rate of network for a faster download, reduces topology un-awareness due to unnecessary traffic and maintains faster download speed compared to other clients [6]. BitTorrent which is a well-known non-localized file sharing software in P2P networking used in common for transferring large file in a vast community environment with an exceptional download speed [7].

The popularity of P2P applications had massive traffic load that revealed doubts about the ability of internet service provider (ISP) that carried P2P traffic [8] and to sustain their cost of transit traffic. Due to these reasons and others inspired research to replace P2P random algorithms with locality-awareness algorithms where locality is a distance measurement method that can be utilized to express locality awareness.

Peer-to-peer (P2P) locality has recently raised lots of interest locally as its written content distribution dramatically raises the traffic within the inter-ISP links, in order to solve this problem the idea to keep a fraction in the P2P site visitors local to help each ISP has been introduced a couple of years ago. Several fundamental issues on locality are being explored such as measuring the content distribution and knowing the harmful effect of locality which intensify the demand of the content file that is shared on the network.P2P applications and ISPs have different lanes of business models that attempt to attract more users by increasing quality of service (QoS).

The fact that allowed P2P application developers to consider underlying networks as free resources and on the other hand ISP's attempting to drag down their inter/intra-domain traffic to increase their profits [9]. This business model authorized ISP's consider P2P applications as a harmful services and thus started to domesticate them by blocking their traffic with the help of shaping devices [5] and on the other side P2P applications counter-strike by encrypting their traffic using port hoping that leads to endless chasing.

However to tackle ISP issues, Autonomous system (AS) hops can be utilized to harvest AS-level to pology information and closely relate the AS-based ISP pricing model. Locality awareness algorithm implementation in P2P application was widely studied in the past years [10]-[12]. Each network on the Internet is recognized by a unique identifier known as Autonomous system number (ASN) which owns a set or a block of Internet Protocol (IP) addresses that have been assigned to it, in order to prevent traffic from propagation, content should be exchanged with other IP addresses in the same AS.

Sniffing is one of the most effective techniques in attacking a wireless network. Sniffer [13] is a program that eavesdrops on the network traffic by grabbing information that travels over a network and the Source for many network-based attacks is passive sniffing. Passive sniffing involves employing a sniffer to be able to monitor these kinds of incoming packets which uses a feature connected with network greeting cards called promiscuous mode. In this mode a network card will pass all packets on the operating structure, rather than those Unicast as well as broadcast towards host [13].

World's prime network protocol analyzer named Wireshark [14] enables to capture and interactively browse the traffic flowing on a computer network. This software is customary across many industries and educational institutions. Wireshark uses a packet capture in short Pcap, an application programming language to capture packets, so it can only capture the packets on the types of networks that Pcap supports. Yi Cui, *et al.*, [15] proposes locality awareness in bit torrent like P2P applications which proposes an optimal solution with minimum AS hop count distribution structure and also describes that seeding cannot improve standard bit torrent download time but can improve its locality policies significantly.

The paper is organized as follows. Section 2 depicts the methodology of achieving goals of our study. Results analysis and performance of Locality and TopBT is studied and discussed in Section 3. Finally, Section 4 gives conclusion and possible future work to improve the quality of service in P2P applications.

2. Methodology and Data Collection

Our methodology of collecting data is to download Torrent files using two different file sharing applications such as Bit Torrent and Top-BT, which were operated in two separate computers. The download time of both Torrent clients was calculated and recorded simultaneously. Wireshark captures and save data packets of both Torrent clients. Then, a utility software was used to extract source and destination IP addresses form Wireshark captured files.

AWK tool [16] was used to delete the duplicate of the source and destination IP addresses. Cymru tool [17] had been utilized by which IP addresses were converted into Autonomous System Numbers (ASN). Java code was developed to find the AS paths from source IP address to destination IP address. By extracting these paths, we then compared the paths generated by the BitTorrent and Top-BT applications. This procedure had been applied on different file formats (Audio, Video, Application files, etc.). Hence the whole data was collected at particular geographical location.

During the downloading of some files we came across a huge download time duration by which a user may lose interest in downloading that particular file. Investigation of the scenario was done to show the impact of locality on the quality of experience. Our calculation was based on the Autonomous System Number and after gathering these numbers, AS paths has been our metric to measure locality.

The following steps were taken to achieve our goal. First we reviewed the concepts of Peer to Peer network (P2P) and Locality revision on the P2P applications. Then, Wireshark and AWK software tools were used for data collection. Finally, Java program and Cymru software tool were employed for data analysis part by which IP addresses were extracted and has been converted to Autonomous System Numbers (ASN).

Two well-known P2P file sharing systems were utilized, namely, Bit Torrent and Top-BT. Software programming and simulation tools such as Java, AWK, Cymru were adapted to map IP's into AS numbers. MS-Excel has been used to show AS paths as final output by which the paths between Bit-torrent and TopBT are compared to measure their QoS and Locality in a P2P network.

3. Results

Locality awareness has emerged as the anchor to tackle the unwanted traffic issue where locality awareness algorithms allow peers to measure their distances from other nodes and utilizes this knowledge in selecting near content sources. To implement this algorithm, many issues must be tackled. For example, how to measure distances? How to find location? How to define near nodes and far nodes?

Collecting underlying network measurements and utilizing this information is the way to answer the previous questions. Peer should have the ability to measure the AS hop count path to reach different peers and must have the ability to map IP addresses into their AS numbers that can be able to measure delay, bandwidth and loss in the path. Finally, they should have algorithms that utilize this information (Locality algorithm). The main objective of locality awareness studies is to construct a P2P system that satisfies the requirement of ISPs by reducing the hops count and increase the number of local source nodes in one way and on the other end there shouldn't be impact on quality of experience in P2P networks.

Our results show that the average AS hops count path between neighbors in TopBT platform is shorter than the distances between neighbors in BT network. In addition, we have observed from our results that locality awareness implemented in TopBT has impact in reducing the intra-domain traffic that passes between AS's. Unfortunately, the implementation of locality awareness algorithm may reduce the performance, Quality of Service (QoS) and Quality of Experience (QoE), of P2P networks if the required content is unpopular.

In other words, the popularity of file in P2P file sharing network may affect implementation of locality awareness algorithm which means that the locality awareness algorithm requires a popularity of files to increase the performance of P2P applications or it will decrease its normal performance. In our measurement study we have obtained our results for Autonomous System paths of sources and destinations on Inter Autonomous System level routing.

In **Figure 1** we have evaluated average Autonomous System paths of Audio files in which TopBT has better download rate than BitTorrent. **Figure 2** and **Figure 3** shows the AS source path comparison of Video files and Application files respectively whereas the Video files that contain large data size results in such case TopBT has performed good as shown in **Figure 2**. Finally, we note that **Figure 4** show Document source average AS path files.

In P2P networks, nodes act as client and server simultaneously. **Figure 5** shows average AS paths of Audio destination files, in **Figure 6** and **Figure 7** average AS path destination files of Video and Application are shown in which the performance of BitTorrent is slight better than TopBT. However TopBT had good performance overall, whereas **Figure 8** shows the average AS hops path of destinations of Document files to compare the performance of TopBT with BitTorrent.

By observing these average AS paths of source files and destinationfiles figures respectively we can list out our findings, At first we noticed average AS hops paths of TopBT is shorter in most of the cases. However, in some cases this path is longer than BitTorrent. The reason is that the downloaded files in these cases are not popular, which means that there are no localized nodes near to download the file from. In this case TopBT attempts

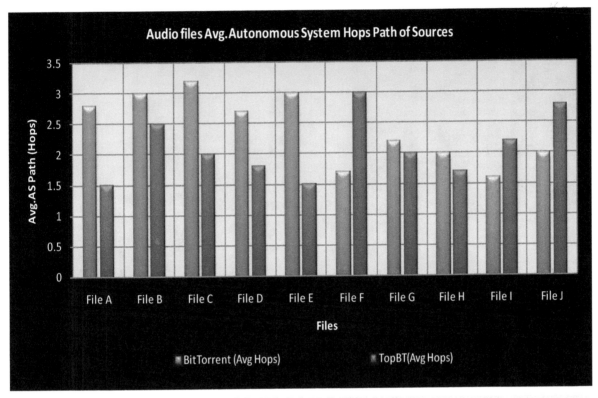

Figure 1. Source autonomous system hops path for audio files.

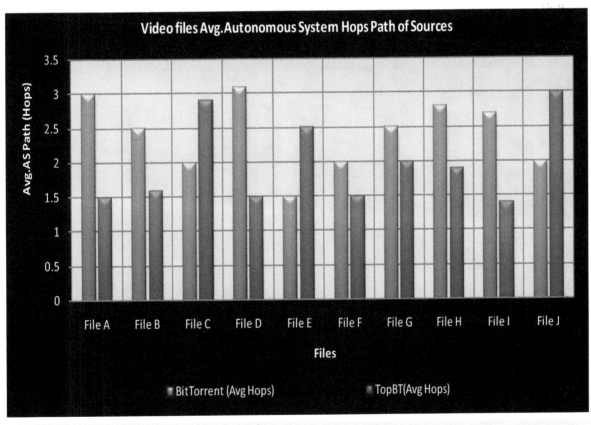

Figure 2. Source autonomous system hops path for Video files.

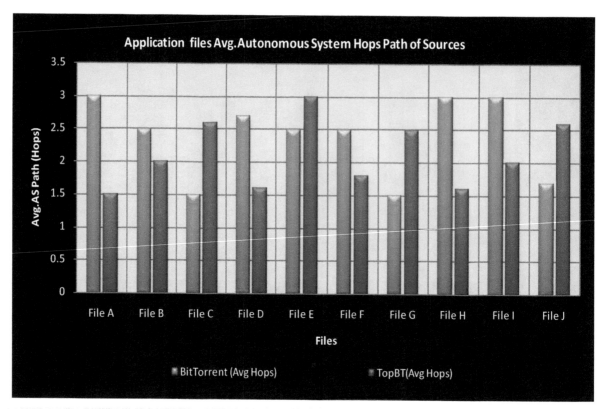

Figure 3. Source autonomous system hops path for application files.

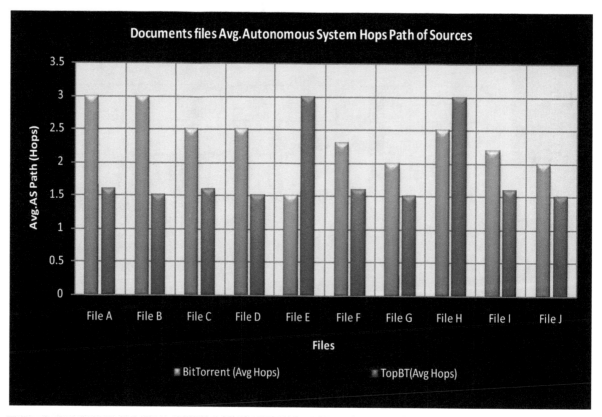

Figure 4. Source autonomous system hops path for documents files.

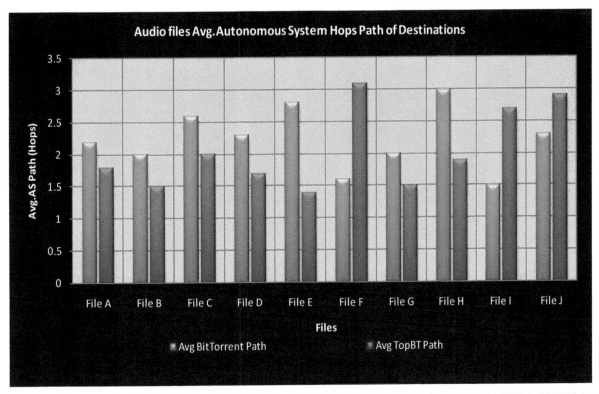

Figure 5. Destination autonomous system hops path for audio files.

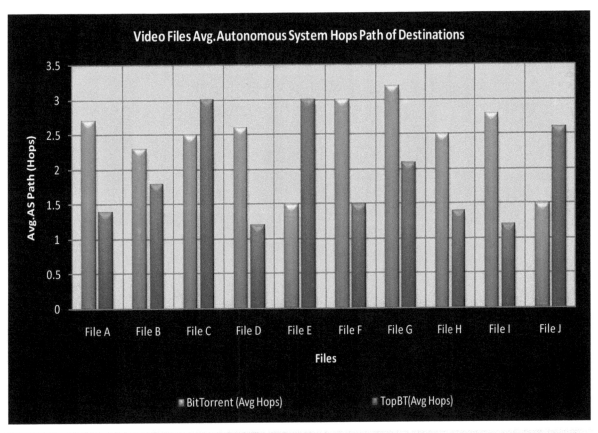

Figure 6. Destination autonomous system hops path for video files.

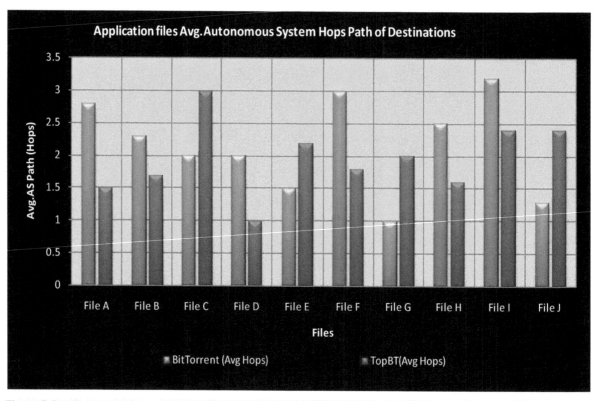

Figure 7. Destination autonomous system hops path for application files.

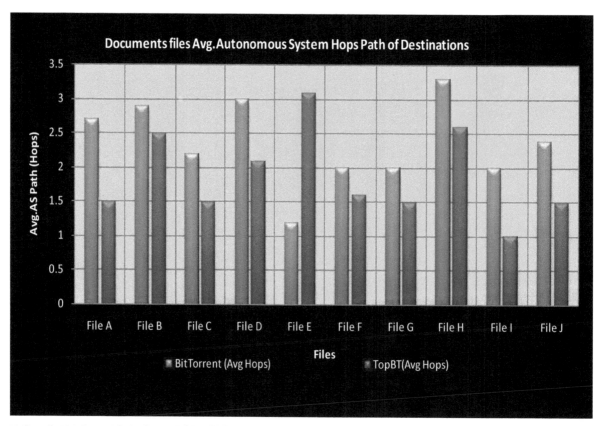

Figure 8. Destination autonomous system hops path for documents files.

to download the files from faster nodes with highest upload bandwidth and hence this fact has increased the path for such scenarios. We can observe that the download time has reduced in many scenarios for TopBT, this shows that locality can also help in improving the Quality of Experience. Unfortunately, this is not the case for all files. TopBT attempts to reduce paths' length first which can affect the download time by downloading the file from closer procrastinate nodes.

4. Conclusions & Future Work

In this work, we conducted a measurement study to investigate the advantages and drawbacks of implementing locality awareness algorithm in P2P networks and examined the locality awareness algorithm in BitTorrent and TopBT. We have compared the performance of TopBT with BitTorrent and utilized Wireshark tool to collect information from P2P network. In addition, an AS graph has been constructed to implement a shortest path algorithm to measure AS hops count between nodes in which collected peers from BitTorrent have been used as input to measure their destinations.

In future work, we can use other P2P applications and compare their results with each other and also investigate this measurement study in different locations and diverse ISP internet connections.

P2P model will remain dominating in coming years as we believe that research and development will continue to adapt P2P overlays which are more suitable for current internet infrastructure. P2P systems evolution will provide insights into the development of other large-scale distributed systems.

References

[1] Quang, H.V., Minhai, L. and Beng, C.O. (2010) Peer to Peer Computing Principles and Applications. New York City. http://dx.doi.org/10.1007/978-3-642-03514-2

[2] Masoud, M.Z., Hei, X.J. and Cheng, W.Q. (2012) Constructing a Locality-Aware ISP-Friendly Peer-to-Peer Live Streaming Architecture. *Proceeding of International Conference on Information Science and Technology (ICIST)*, Hubei, 23-25 March 2012, 368-376. http://dx.doi.org/10.1109/ICIST.2012.6221670

[3] Alhaisoni, M., Liotta, A. and Ghanbari, M. (2009) Performance Analysis and Evaluation of P2PTV Streaming Behavior. *International Symposium on Computers and Communications (ISCC)*, Sousse, 89-94. http://dx.doi.org/10.1109/ISCC.2009.5202402

[4] Zhang, D.Y., Zheng, C., Zhang, H.L. and Yu, H.L. (2010) Identification and Analysis of Skype Peer-to-Peer Traffic. *International Conference on Internet and Web Applications and Services (ICIW)*, Barcelona, 200-206. http://dx.doi.org/10.1109/ICIW.2010.36

[5] Aggarwal, V., Akonjang, O. and Feldmann, A. (2008) Improving User and ISP Experience through ISP-Aided P2P Locality. *Proceedings of INFOCOM Workshops*, Phoenix, 1-6. http://dx.doi.org/10.1109/INFOCOM.2008.4544640

[6] Ren, S., Tan, E., Luo, T., Guo, L., Chen, S. and Zhang, X. (2010) TopBT: A Topology-Aware and Infrastructure-Independent BitTorrent Client. *Proceedings of INFOCOM*, San Diego, 1-9. http://dx.doi.org/10.1109/INFCOM.2010.5461969

[7] Xia, R.L. and Muppala, J.K. (2010) A Survey of BitTorrent Performance. *International Journal of Communications Surveys & Tutorials*, **12**, 140-158. http://dx.doi.org/10.1109/SURV.2010.021110.00036

[8] Fras, M., Klampfer, S. and Cucej, Z. (2008) Impact of P2P Traffic on IP Communication Network Performances. *Proceeding of International Conference on Systems, Signals and Image Processing*, Bratislava, 205-208. http://dx.doi.org/10.1109/IWSSIP.2008.4604403

[9] Blond, S.L., Legout, A. and Dabbous, W. (2004) Pushing BitTorrent Locality to the Limit. *International Journal of Computer and Telecommunications Networking*, **55**, 541-557. http://dx.doi.org/10.1016/j.comnet.2010.09.014

[10] Coffins, D.R. and Bustamante, F.E. (2008) Taming the Torrent: A Practical Approach to Reducing Cross-ISP Traffic in Peer-to-Peer Systems. *Proceeding of ACM SIGCOMM Conference on Data Communication*, 363-374. http://dx.doi.org/10.1145/1402958.1403000

[11] Lewis, P.R., *et al.* (2011) A Survey of Self-Awareness and Its Application in Computing Systems. *Proceeding of International Conference on Self-Adaptive and Self-Organizing Systems Workshops (ICSASOW)*, Ann Arbor, 102-107. http://dx.doi.org/10.1109/SASOW.2011.25

[12] Agarwal, V., Feldmann, A. and Scheideler, C. (2007) Can ISPs and P2P Users Cooperate for Improved Performance. *ACM SIGCOMM Computer Communication Review Journal*, **37**, 29-40.

[13] Chomsiri, T. (2008) Sniffing Packets on LAN without ARP Spoofing. *Proceeding of International Conference on Convergence and Hybrid Information Technology (ICCIT)*, **2**, 472-477. http://dx.doi.org/10.1109/ICCIT.2008.318

[14] Wang, S.Q., Xu, D.S. and Yan, S.L. (2010) Analysis and Application of Wireshark in TCP/IP Protocol Teaching. *Proceeding of International Conference on E-Health Networking Digital Ecosystems and Technologies* (*EDT*), **2**, 269-272. http://dx.doi.org/10.1109/EDT.2010.5496372

[15] Liu, B., Cui, Y., Lu, Y.S. and Xue, Y. (2009) Locality-Awareness in BitTorrent-Like P2P Applications. *IEEE Transactions on Multimedia—Special Section on Communities and Media Computing Journal*, **11**, 361-371. http://dx.doi.org/10.1109/TMM.2009.2012911

[16] Aho, A.V., Kernighan, B.W. and Weinberger, P.J. (1988) The AWK Programming Language. New York.

[17] Cymru. http://www.team-cymru.org/Services/ip-to-asn.html

Permissions

The contributors of this book come from diverse backgrounds, making this book a truly international effort. This book will bring forth new frontiers with its revolutionizing research information and detailed analysis of the nascent developments around the world.

We would like to thank all the contributing authors for lending their expertise to make the book truly unique. They have played a crucial role in the development of this book. Without their invaluable contributions this book wouldn't have been possible. They have made vital efforts to compile up to date information on the varied aspects of this subject to make this book a valuable addition to the collection of many professionals and students.

This book was conceptualized with the vision of imparting up-to-date information and advanced data in this field. To ensure the same, a matchless editorial board was set up. Every individual on the board went through rigorous rounds of assessment to prove their worth. After which they invested a large part of their time researching and compiling the most relevant data for our readers.

The editorial board has been involved in producing this book since its inception. They have spent rigorous hours researching and exploring the diverse topics which have resulted in the successful publishing of this book. They have passed on their knowledge of decades through this book. To expedite this challenging task, the publisher supported the team at every step. A small team of assistant editors was also appointed to further simplify the editing procedure and attain best results for the readers.

Apart from the editorial board, the designing team has also invested a significant amount of their time in understanding the subject and creating the most relevant covers. They scrutinized every image to scout for the most suitable representation of the subject and create an appropriate cover for the book.

The publishing team has been an ardent support to the editorial, designing and production team. Their endless efforts to recruit the best for this project, has resulted in the accomplishment of this book. They are a veteran in the field of academics and their pool of knowledge is as vast as their experience in printing. Their expertise and guidance has proved useful at every step. Their uncompromising quality standards have made this book an exceptional effort. Their encouragement from time to time has been an inspiration for everyone.

The publisher and the editorial board hope that this book will prove to be a valuable piece of knowledge for researchers, students, practitioners and scholars across the globe.

List of Contributors

Akinobu Nemoto
Department of Medical Informatics, Yokohama City University, Yokohama, Japan

Pham Thanh Hiep and Ryuji Kohno
Division of Physics, Electrical and Computer Engineering, Yokohama National University, Yokohama, Japan

Omar Daoud
Communications and Electronics Engineering Department, Philadelphia University, Amman, Jordan

Yasusi Kanada
Central Research Laboratory, Hitachi, Ltd., Yokohama, Japan

Faisal Shahzad, Muhammad Faheem Mushtaq, Saleem Ullah, Shahzada Khurram and Najia Saher
Department of Computer Science & IT, The Islamia University of Bahawalpur, Bahawalpur, Pakistan

M. Abubakar Siddique
College of Computer Science, Chongqing University, Chongqing, China

Jinbao Zhang, Song Chen and Qing He
School of Electronic and Information Engineering, Beijing Jiaotong University, Beijing, China
Beijing Engineering Research Center of EMC and GNSS Technology for Rail Transportation, Beijing, China

Nusaibah M. Al-Ratta, Mznah Al-Rodhaan and Abdullah Al-Dhelaan
Computer Science Department, College of Computer and Information Sciences, King Saud University, Saudi Arabia

Anis Zarrad and Ahmed Redha Mahlous
Department of Computer Science and Information Systems, Prince Sultan University, Riyadh, Saudi Arabia

Tatsuya Yamada, Mayu Mitsukawa, Hideki Shimada and Kenya Sato
Mobility Research Center, Doshisha University, Kyoto, Japan

Muthana Najim Abdulleh and Salman Yussof
College of Information Technology, Universiti Tenaga Nasional, Kajang, Malaysia

Hothefa Shaker Jassim
College of Engineering, Komar University of Science and Technology, Sulaymaniyah, Iraq

Ahmed Redha Mahlous
Department of Computer and Information Sciences, Prince Sultan University, Riyadh, Saudi Arabia

Maria Catarina Taful and João Pires
Department of Electrical and Computer Engineering, Instituto Superior Técnico, University of Lisbon, Lisbon, Portugal

João Pires
Institute of Telecommunications, Instituto Superior Técnico, University of Lisbon, Lisbon, Portugal

Muhammad Umar Aftab
Department of Computer Science and Engineering, Yuan Ze University, Taiwan

Omair Ashraf
Department of Computer Science, NFC-IEFR, Faisalabad, Pakistan

Muhammad Irfan
Land Record Management and Information System, Faisalabad, Pakistan

Muhammad Majid
IT Department, Post Graduate College, Faisalabad, Pakistan

Amna Nisar, Muhammad Asif Habib and Muhammad Umar Aftab
Department of Computer Science, National Textile University, Faisalabad, Pakistan

Hirokazu Miura and Hirokazu Taki
Faculty of Systems Engineering, Wakayama University, Wakayama, Japan

Hirokazu Miura and Hideki Tode
Department of Computer Science and Intelligent Systems, Osaka Prefecture University, Sakai, Japan

Mohsen M. Tantawy
National Telecommunication Institute (NTI), Cairo, Egypt

Mahmoud Maqableh and Ra'ed Moh'd Taisir Masa'deh
Management Information Systems, Faculty of Business, The University of Jordan, Amman, Jordan

Lama Rajab, Walaa Quteshat, Tahani Khatib and Huda Karajeh
Computer Information Systems, King Abdullah II School for Information Technology, University of Jordan, Amman, Jordan

Mohammed Younus Talha, Rabah W. Aldhaheri and Mohammad H. Awedh
Department of Electrical and Computer Engineering, King Abdulaziz University, Jeddah, Saudi Arabia

Printed in the USA
CPSIA information can be obtained
at www.ICGtesting.com
JSHW051445221024
72173JS00006B/1588

9 781632 405470